CW01512529

DARWIN AND WOMEN

Darwin and women is a publication of the Darwin Correspondence Project. The Project was founded in 1974 with the aim of publishing all known letters to and from Charles Darwin. The editors of the project have produced a number of special publications in addition to the main series. This book focuses on Darwin's correspondence with women and on the lives of the women he knew and wrote to. It includes a large number of hitherto unpublished letters between members of Darwin's family and their friends, letters that will not be published in the main series but that throw light on the lives of the women of his circle. The letters included are by turns entertaining, intriguing and challenging, and are organised into thematic chapters that set them in an accessible narrative context. Darwin's famous remarks on women's intelligence in *Descent of man* provide a recurring motif, and are discussed in the foreword (by Dame Gillian Beer), and in the introduction.

Contributors:

DR SAMANTHA EVANS is an associate editor of the Darwin Correspondence Project.

DR CHARISSA VARMA has collaborated for many years with the Darwin Correspondence Project, most recently as a SSHRC Post Doctoral Fellow. She is an affiliated scholar in the Department of History and Philosophy of Science at the University of Cambridge and a research associate at Darwin College.

DR PAUL WHITE is an associate editor of the Darwin Correspondence Project and an affiliated scholar in the Department of History and Philosophy of Science at the University of Cambridge.

Support for editing the *Correspondence of Charles Darwin* has been received from the Alfred P. Sloan Foundation, the Andrew W. Mellon Foundation, the Arts and Humanities Research Council, the British Academy, the British Ecological Society, the Evolution Education Trust, the Isaac Newton Trust, the John Templeton Foundation, the National Endowment for the Humanities, the National Science Foundation, the Natural Environment Research Council, the Royal Society of London, the Stifterverband für die Deutsche Wissenschaft and the Wellcome Trust. The National Endowment for the Humanities funding of the work was under grants nos. RE-23166-75-513, RE-27067-77-1359, RE-0082-80-1628, RE-20166-82, RE-20480-85, RE-20764-89, RE-20913-91, RE-21097-93, RE-21282-95, RZ-20018-97, RZ-20393-99, RZ-20849-02 and RQ-50388-09; the National Science Foundation funding of the work was under grants nos. SOC-75-15840, SOC-76-82775, SES-7912492, SES-8517189, SBR-9020874, SBR-9616619, SES-0135528, SES-0646230 and SES-0957520. Any opinions, findings, conclusions or recommendations expressed in this publication are those of the editors and do not necessarily reflect the views of the grantors.

Other publications of the Darwin Correspondence Project

The correspondence of Charles Darwin, edited by Frederick Burkhardt *et al.* 24 vols. to date.

Origins: selected letters of Charles Darwin, 1822-1859, edited by Frederick Burkhardt.

Evolution: selected letters of Charles Darwin, 1860-1870, edited by Frederick Burkhardt, Samantha Evans and Alison M. Pearn.

Charles Darwin: the Beagle letters, edited by Frederick Burkhardt *et al.*

A voyage round the world: Charles Darwin and the Beagle collections in the University of Cambridge, edited by Alison M. Pearn.

DARWIN AND WOMEN

A Selection of Letters

BY SAMANTHA EVANS

Foreword by Dame Gillian Beer

CAMBRIDGE
UNIVERSITY PRESS

CAMBRIDGE
UNIVERSITY PRESS

University Printing House, Cambridge CB2 8BS, United Kingdom

One Liberty Plaza, 20th Floor, New York, NY 10006, USA

477 Williamstown Road, Port Melbourne, VIC 3207, Australia

4843/24, 2nd Floor, Ansari Road, Daryaganj, Delhi – 110002, India

79 Anson Road, #06–04/06, Singapore 079906

Cambridge University Press is part of the University of Cambridge.

It furthers the University's mission by disseminating knowledge in the pursuit of education, learning, and research at the highest international levels of excellence.

www.cambridge.org
Information on this title: www.cambridge.org/9781107158863
DOI: 10.1017/9781316670033

First published 2017

Printed in the United Kingdom by Clays, St Ives plc

A catalogue record for this publication is available from the British Library.

Library of Congress Cataloging-in-Publication Data

Names: Darwin, Charles, 1809-1882, author. | Evans, Samantha, compiler.
Title: Darwin and women : a selection of letters / [compiled by] by Samantha Evans ; foreword by Dame Gillian Beer.
Description: New York : Cambridge University Press, 2017. | Includes bibliographical references and index.
Identifiers: LCCN 2016046824 | ISBN 9781107158863
Subjects: LCSH: Darwin, Charles, 1809-1882–Correspondence. | Naturalists–England–Correspondence. | Women.
Classification: LCC QH31.D2 D3625 2017 | DDC 576.8/2–dc23
LC record available at https://lccn.loc.gov/2016046824

ISBN 978-1-107-15886-3 Hardback

Contents

Illustrations

Foreword
by Dame Gillian Beer

Darwin spent his life surrounded by spirited, enlightened, and supportive women, many of whom were actively involved in scientific enterprises. He relished female company and appreciated the precise observations, assiduous collecting of evidence, and the firm stand on principles of many of his female acquaintance. Yet he distinguished female capacities from those of human males in a deterministic and somewhat demeaning manner in the *Descent of Man*. The later part of this foreword will address that puzzling paradox and consider some of the pressures that go into it.

First, though, this cornucopia of correspondence demonstrates how throughout his adult life Darwin was in touch with an array of intelligent women. The letters from the Darwin Correspondence are here revealingly augmented with hitherto unpublished family letters in which 'Charles' or 'F.' (for Father) and his concerns are part of the network of preoccupations shared by very different people. Wherever you turn in this volume there are insights into the workings of the scientific community, often cast in the guise of acquaintance and gossip. The economics of funding research, the pressures of gaining a livelihood and sustaining a career, the innovative exchanges between friends and colleagues, are here understood anew as they become filtered through the experience of women, often acting as unpaid assistants, or translators, and research companions to their husbands, as was the case with Mary Lyell or Ellen Lubbock, for example. Indeed, Emma Darwin translated for Charles from French, German, and Italian and read much of his work as he proceeded. The particular significance of explicit, and implicit, exchanges sometimes emerges gradually through the organisation of the present volume, which is in terms of themes and clusters. These range from 'Marriage' to 'Companion Animals' and 'Religion' by way of 'Children', 'Scientific Wives and Allies', 'Travellers', 'Servants and Governesses', to the 'Ascent of Woman', with editors, and observing plants and humans, and other important topics on the way. One outcome of this arrangement is that people crop up under very different headings so that we gradually come to see the different roles they play. The arrangement is not always chronological and the reader needs to be aware of that. The editors' commentary creates a narrative and contextualising thread. The letters range from courteous formal expressions of gratitude to strangers to intimate and sometimes harrowing accounts of family events. This latter is particularly true for the chapter on children.

Children

Births, marriages, and deaths shape every family but for Victorian house-holds that trinity could quite often occur in a more tragic order: marriages, births, and deaths. Childbirth and child-rearing were beset with dangers, first for the mother and baby and then throughout childhood from dangerous infections and illnesses. Perhaps inevitably then, the collection of letters in this chapter makes for painful reading. The death of ten-year-old Annie in 1850 has often been movingly written about and here we learn more of the impact of that death on other members of the household and the deep distress of her nurse, Jessie Brodie, and her governess, Catherine Thorley.

Crises demand correspondence and the letters around the childbirth and death in 1876 of Amy Ruck, Francis (Frank) Darwin's young wife, take the reader close into the raw grief of both families. To their son William, Emma vividly sketches some of the different reactions: Darwin 'distracting his mind with schemes about building an additional room so that Frank may be made comfortable', Bessy, their shy youngest daughter, who 'can only sleep very little and is utterly shattered':

> She feels truly that she can never hope to have the loss of Amy replaced. She was so sympathetic & the only person B. could be open with. (41)

The last paragraph of Emma's letter is challenging in its matter-of-fact insight:

> My heart aches whenever I think of Frank; but now he is out of our sight we shall be able to forget him more & take to our usual occupations. (41)

The surviving baby, Bernard, was their first grandson and, as it turned out, he and his father lived with Charles and Emma after Amy's death.

Indeed, it strikes me as very probable that it was the presence of this baby in the house that awakened Darwin's special interest in the discussions of child development current in *Mind*[1] in 1877. In 'A Biographical Sketch of an Infant' he went back, poignantly, to the diary he had kept of William's infancy and toddlerdom thirty-seven years earlier and added to it some observations from his later theories. In an incautious couple of sentences, and based it has to be said on a very narrow observational cohort, he drew some tentative conclusions about the physical capacities of male and female children:

> When two years and three months, he became a great adept at throwing books or sticks, &c., at anyone who offended him; and so it was with some of my other sons. On the other hand, I could never see

[1] *Mind: a Quarterly Review of Psychology and Philosophy* 2 (July 1877): 285–94.

a trace of such aptitude in my infant daughters; and this makes me think that a tendency to throw objects is inherited by boys. (p. 288)

This example suggests that Darwin could be somewhat cavalier with the evidence in attributing inherited tendencies to the sexes!

He was certainly a devoted and observant father and the notes that he and Emma made on the sayings and behaviour of their children, gathered as Appendix III to volume 4 of the *Correspondence*, would leaven the atmosphere of sorrow and crisis that prevails in much of this chapter despite the editors' account of the usually happy atmosphere at Down House.[2] Lenny, in particular, had a way with words:

Lenny "Is the sky a sort of nowhere?
L. I've opened the window an atomist bit.
looking thro' that red thing unbetters me. (looking thro' a bit of red glass at the garden) (46v.)
Lenny,— "It sometimes happens that I am happy"
April 10th: In the morning whilst I was shaving, Lenny kept talking to me so I said, "Lenny I cannot talk while I am shaving"— "But you can talk, when you are *unshaving*".— (50v, 50bv.)

Not all family life can find its way into letters, particularly not the precious humdrum of everyday.

Gossip and Politics

The collection gives us insight into the workings of acquaintance and exclusion within the scientific community of Darwin's time. For example, a letter from J. D. Hooker to Darwin in 1865 comments extensively on the fact that 'Lady Lyell will not call on Mrs. Busk nor invite the Busks to her parties' (52). What might seem a matter of trivial personal distaste is revealed by the editor's commentary to have wider ramifications. The Busks were a distinguished scientific pair who could ordinarily have expected to be invited to a semi-public gathering of scientists of this kind. The reasons for their exclusion seem to have been, the editor comments, that 'Ellen Busk was known as a freethinker and religious sceptic' (52). The party attendance or exclusion carried questions about religion, evolutionary debate, and even plagiarism.

Some of those same questions emerge again around the Busks in a different setting: the controversy over the merging of the Ethnological Society, originally an offshoot of the Aborigines' Protection Society and relatively liberal in outlook, with the Anthropological Society, a generally anti-Darwinian society, in 1871. Ellen Lubbock, who herself practised as an archaeologist, writes vigorously to Emma Darwin, in tones that take us

2 *The Correspondence of Charles Darwin*, volume 4 1847–50 (Cambridge: Cambridge University Press, 1988), Appendix III, 'Darwin's Observations on His Children', pp. 427, 429.

close into the exasperation and urgency felt by the participants, and yet she manages to keep a persuasive, even frolicsome, lightness. The letter is clearly intended to be seen by Charles Darwin:

> I hate begging—so now you will perceive I am going to beg. Yesterday I was at the Busks', & Mʳ Busk was groaning & lamenting over his Presidency of the Anthro—(I never can spell the horrid word)—Society—the name irritates him, as it does John, & it *isn't* the right one. We never wanted to be merged & swallowed whole in and by this mushroom society, with no good men in it— So I said well, why not alter it back to the Ethnological, which was the first & real root of the thing? To which he replied despondently that they were in debt £700.

The conflicted early history of anthropology is all caught into this letter with the stresses and strains of knowledge ownership and interpretation hinted at and sometimes breaking out: 'Poor old Mʳ Crawford would have given every penny he had, in fact I should think he turned in his grave when his pet Society was named after his bitterest enemies.' The tribalism of Victorian scientific society attempting new kinds of knowledge is hinted at in these charmingly wheedling words, as Ellen Lubbock (half) fears being thought 'very meddlesome' (58).

A tone of playful banter is frequent in the correspondence of Darwin's close female acquaintance. Henrietta Huxley teases him about his disparagement of a line from Tennyson ('And he meant, he said he meant, Perhaps he meant, or partly meant you well.')

> In the first place it was very mean of you to give the lines without the context shockingly Owenlike (7)

Darwin had suffered from Richard Owen's maliciously taking his statements out of their context. Moreover, she points out, he has got the source of the quotation wrong. It is from 'Sea Dreams' not 'Enoch Arden' and he has thus 'damaged [his] reputation for accuracy'.

> If the "facts?!" in the Origin of Species are of this sort—I agree with the Bishop of Oxford [another of Darwin's adversaries].

This is breathtakingly impertinent, as only the best of friends can be, and it shows how his friends could rely on Darwin's sense of humour to accept such sallies. Intriguingly, he found it harder to accept his family's suggestions and corrections to his work, as Emma remarked. Henrietta became the exception to this as she more and more took on the role of editor for his work, especially of *Descent of Man*. He much valued the work she did, as an 1871 letter from him to her makes clear:

> Several reviewers speak of the lucid vigorous style etc.— Now I know how much I owe to you in this respect, which includes arrangement,

not to mention still more important aids in the reasoning. …
Goodbye my very dear coadjutor & fellow-labourer
Your affec^ate. father. Ch Darwin (143)

Henrietta was paid for her work by her father though 'as a memorial or souvenir rather than as wages.' (143)

Frances Power Cobbe, one of the most dynamic of his correspondents, and an activist in an array of campaigns, particularly anti-vivisectionism and rational dress, in a letter probably from 1870 urges Darwin to take Kant seriously and to unite his own lines of thought and 'let us see how metaphysics & physics form one great philosophy' (156). Perhaps to soften the force of her criticism she then uses a somewhat winsome apology that declares and jokes about her own feminism:

> Pray forgive dear Mr Darwin, my infinite impudence! Though I attended on Saturday a most successful Woman's Rights Meeting I am of opinion that our Ancient privilege of talking nonsense even to those we most deeply honour, is one not to be parted with on any terms! (156)

The hidden political activity of women within the scientific community becomes manifest in the pages of this volume, from Arabella Buckley organising the campaign to procure a government pension for Alfred Russel Wallace to friends rallying to help Thomas Henry Huxley. But despite all the backstage activity and the drawing on funds that may well have been her own before marriage it was sometimes felt to be distasteful for a woman to appear openly on the list as a donor herself.

Many of the women who operated in these quiet political ways were themselves skilled scientific observers and collectors, and there were also many who acted as Darwin's scientific informants without ever meeting him face to face, providing him with crucial material particularly in the area of botany.

Poets Reply

Darwin's presence reached out to affect other creative women, within his lifetime and just after his death, particularly the poets of the age such as Mathilde Blind, whose *Ascent of Man* (1888) challenges the exclusionary aspects of natural selection and chooses to concentrate on the less enabled as well as on the wild drama of initial creation. May Kendall and Constance Naden both explore in light verse the implicit hierarchies lodged in evolutionary ideas and challenge its tendency to promote progress in such a way that the pride of humankind is left intact. These are writers who are, one may say, indirectly corresponding with Darwin.

In her jaunty and satiric poem 'Solomon Redivivus, 1886', Constance Naden imagines King Solomon reappearing in 1886 as 'the modern Sage'.

Naden first describes this 'two sexes united in the same individual' and then the whole sequence of evolution by which the sexes were separated. Naden, herself well educated scientifically, considers in the poem the various phases of evolutionary development. She seems indeed to have had in mind a particular summary paragraph from the *Descent*:

> The Quadrumana and all the higher mammals are probably derived from an ancient marsupial animal, and this through a long line of diversified forms, from some amphibian-like creature, and this again from some fish-like animal. In the dim obscurity of the past we can see that the early progenitor of all the Vertebrata must have been an aquatic animal, provided with branchiae, with the two sexes united in the same individual, and with the most important organs of the body (such as the brain and heart) imperfectly or not at all developed. (2nd ed., p. 609)

... 'the two sexes united in the same individual' and no developed heart or brain—Naden seizes on that description.

She writes in the voice of the returned King Solomon, again wooing and addressing the Queen of Sheba. I quote here a few stanzas from a three-page poem.

We were a soft Amoeba
  In ages past and gone,
Ere you were Queen of Sheba,
  And I King Solomon.

Unorganed, undivided,
  We lived in happy sloth,
And all that you did I did,
  One dinner nourished both:

Till you incurred the odium
  Of fission and divorce—
A severed pseudopodium
  You strayed your lonely course.

So through the form of fish, reptile, mammal, at last appears our predecessor, the naked ape:

But now, disdaining trammels
  Of scale and limbless coil,
Through every grade of mammals
  We passed with upward toil.

Till, anthropoid and wary
  Appeared the parent ape,

And soon we grew less hairy,
  And soon began to drape.

So, from that soft Amoeba
  In ages past and gone,
You've grown the Queen of Sheba,
  And I King Solomon.[3]

Naden's poem is both a neat lesson in evolutionary sequence and a reminder of how the diverse forms of life cling still within the present form of the human. Her use of 'we' represents not only the loving pair of king and queen but all the beings through whom they emerged to their present state. 'We' are ascidian, fish, and reptile, as well as the self-crowned height of evolutionary history. And she picks up Darwin's long discussion of hairy or hairless men in which he, perhaps unwisely, takes for granted the beauty of the beard in the opinion of his readers. By taking on the voice of King Solomon, Naden both mocks and relishes the claims of men: it is the Queen of Sheba who precipitated divorce and development by 'fission' from him, and from that primordial state when 'Whatever you did I did, / One dinner nourished both.' Naden is intrigued by the idea of the single progenitor but also feels glee at the disruptive powers of the female who asserts her independence. She draws inspiration for her satire quite specifically from the *Descent*.

Not all these women poets knew or corresponded directly with Darwin but one who did was Emily Pfeiffer. Pfeiffer wrote about the situation of women in essays in the journals (156) and her poetry has a tonic intensity in its insights into some of the less sanguine aspects of Darwinian theory. The one surviving letter from her to him detaches the idea of beauty from that of fascination in understanding sexual selection: fascination may be malign, beauty is life-giving (157). Darwin himself grew uneasy with the personification of Nature, with its suggested undertow of maternal care, as we can see from the variorum edition of the *Origin*, where he struggles to de-personalise nature and reduce it to law-like processes.

So again it is difficult to avoid personifying the word Nature; but I mean by Nature, only the aggregate action and product of many natural laws, and by laws the sequence of events as ascertained by us. With a little familiarity such superficial objections will be forgotten.[4]

Pfeiffer saw how ill the language of a personified Mother Nature fitted with the implications of Darwin's theory. She wrote several powerful sonnets on the implications of Darwinian evolution: this one addresses the nature of Nature:

---

3 Constance Naden, *Complete Poetical Works*, with an explanatory foreword by Robert Lewins (London: Bickers & Son, 1894), pp. 317–19.

4 Morse Peckham, *The Origin of Species by Charles Darwin: a variorum text* (Philadelphia: University of Pennsylvania Press, 1959), p. 165.

Dread Force, in whom of old we loved to see
A nursing mother, clothing with her life
   The seeds of Love divine,—with what sore strife
We hold or yield our thoughts of Love and thee!
Thou art not 'Calm,' but restless as the ocean,
   Filling with aimless toil the endless years—
Stumbling on thought, and throwing off the spheres,
Churning the Universe with mindless motion.

Dull fount of joy, unhallowed source of tears,
   Cold motor of our fervid faith and song,
Dead, but engendering life, love, pangs, and fears,
Thou crownedst thy wild work with foulest wrong
When first thou lightedst on a seeming goal,
And darkly blundered on man's suffering soul. (p. 30)

This grim apostrophe to a repudiated mother who proves to be mere machine, Nature, 'churning the Universe with mindless motion', expresses also the struggle in the poet's mind to 'hold or yield our thoughts of Love and thee'. Hers is another kind of personification, not 'only the aggregate action and product of many natural laws, and by laws the sequence of events as ascertained by us'. But Darwin also found it more or less impossible in his descriptions to hold to that reductionist 'only': 'only the aggregate action'. Elsewhere in the same sequence of sonnets Pfeiffer addresses Evolution as 'Hunger': 'Sacred disquietude, divine unrest!':

Hunger that strivest in the restless arms
Of the sea-flower, that drivest rooted things
To break their moorings, that unfoldest wings
In creatures to be rapt above thy harms;

That poem ends:

Thou art the Unknown God on whom we wait:
Thy path the course of our unfolding fate. (p. 51)

The blanking out of knowledge and its substitution by *energy*, an energy that informs all and drives us to seek its meaning, is Pfeiffer's persistent dilemma—and the originality of her insight. Pfeiffer's sonnets—pithy, compressed, yet extreme—refuse to bring Nature back into meaning, even while she longs for a sestet that will restore harmony. She sets severe limits to sympathy. Sympathy does not reach the human from other forms. Instead the task of the human is to reach towards other life forms that *do not* reciprocate. Pfeiffer confronts the degree to which humans share the impersonal drives of all living organisms without being able to find any community with them.

It is perhaps no wonder that Mark Pattison was divided between dismay and admiration in his response, published in a preface to the second edition of her work:

> I think the most striking and original of your sonnets are those inspired by the evolutional idea—an idea or form of universal apprehension, which, like a boa, has infolded all mind in this generation in its inexorable coil. Try as we may, we cannot extricate our thoughts from this serpent's fold. Its pressure upon the soul forces our spirit to cry out with a Laocoon shriek; but though the inspiration of despair, it is inspiration, and poetry is its natural vent.[5]

Pfeiffer articulated troubles that few of Darwin's correspondents could directly address in their letters and her poems expand the effects of his enquiry.

Women's Capacities

Darwin's early life-experience moved from a childhood where sisters were all-important and a youth with many young women friends, to five years in his twenties spent on, and off, the *Beagle*, where his habitual cohort was entirely other men. He encountered women during his land journeys from the ship but these were, so far as we can tell, either formal social encounters with upper-class Latin-American women or cautious observation of indigenous women. His mother had died when he was eight years old and he deeply regretted that he held few memories of her. On his return to London he re-entered the large social network of his extended family and he married his first cousin, Emma Wedgwood, from a lively and enlightened family. The Darwins' own immediate family came to include three living daughters, among six sons, and their household extended to governesses and nurses and female servants. His life after marriage was settled, though he travelled for his health, and occasionally for leisure, within Britain. Because of his very uncertain health, the Darwins did not entertain at large. Close friends, particularly married couples and scientific colleagues like the Huxleys, Lyells, and Lubbocks came to see him, but much of his friendship and acquaintance took place as correspondence.

During his five years of world travel his sisters kept him in touch through their letters with British home life and with them he shared family gossip, literature, and current affairs (though letters had a drag of several months on their arrival). The thrill of constant discovery during that time provided him with a template of achievement entirely occupied by men. However, his intellectual life continued to be profound and wide-ranging throughout the years of later domesticity and his children became his assistants without discrimination of gender. What is striking in this pattern is that the most

---

5 Emily Pfeiffer, *Sonnets*, revised and enlarged edition (London: Field & Tuer, The Leadenhall Press, [1886?]), p. iii.

innovative and adventurous period of his active life, the years on the *Beagle* voyage, left women in the past or in a now impossible future as a clergyman with a little wife. Perhaps a nostalgia for those active male-centred years also inclined him, unawares, to take a lower estimate of women as essentially bound to domestic life.

His daughters were born too soon to enjoy a university education but Henrietta's letters show her wit and authority and she travelled widely in Europe. Annie, who died at ten years old, was felt by all the family to be an unusually gifted child, and his youngest daughter, Elizabeth or Bessie, despite her reclusive reputation, had the enterprise to request to be sent to boarding school as a young girl and attended lectures at University College, London. Emma, his wife, oversaw the large Darwin household, read and wrote letters for him, kept up contacts with other scientific families, played the piano with great proficiency, translated from several languages, read novels with him daily, and bore him many children with all the dangers by which childbirth and child-rearing were then surrounded. Without Emma's presence Darwin's professional achievements would have been hard to solidify. The problems of a scientific household without a competent wife to manage it can be seen in these pages in the sad case of J. D. Hooker, whose domestic confusion and troubles are retailed to Darwin over a number of years. Darwin was also the beneficiary of his father's shrewd investments as well as a skilled manager of financial affairs himself. Neither the women nor the men by whom he was surrounded matched Darwin's achievements: that was not to be discriminated by gender.

When Darwin in the *Descent* commented that women would never equal men until they became the breadwinners he seems not to have noticed that his wife's extraordinary time-management smoothed his uninterrupted researches and writing and thus underpinned the 'breadwinning' of the household. He repeats this view in his late correspondence with Caroline Kennard: 'women must become as regular "bread-winners" as are men' to avoid 'the laws of inheritance'. But then he demurs further: 'we may suspect that the early education of our children, not to mention the happiness of our homes, would in this case greatly suffer' (226). Thus women are corralled within an argumentative loop: they cannot catch up with men until they are active in the wider world but if they are so active, they will lose the innate moral superiority with which he at present endows them, because they will be obliged to sacrifice their families to their ambitions. Yet in his life experience he had encountered, and appreciated, both as correspondents and as personal acquaintances, many women who succeeded in balancing scientific fieldwork with more domestic commitments. What few women then held was independent wealth and access to higher education.

Another reason, perhaps, for Darwin's difficulty in recognising female capacities was the inclination of the language that he was working with. Until the 1870s Darwin did not publish extensively about human beings and their descent or liaisons. Then in quick succession he published *The Descent of Man, and Selection in Relation to Sex* (1871) and *The Expression of the Emotions in Man and Animals* (1872). That second work was originally to have formed part of the *Descent* but it grew too large. The *Descent* itself is on an enormous scale and explores the issue of sexual selection in ways that demanded quite new thinking from Darwin, though it had been touched on in the *Origin*. The sexual behaviour of different human groups is studied in the *Descent* alongside that of other kinds, as also are the physical differences between sexes in a range of creatures. And here we begin to see the problem that Darwin has not so much introduced as illuminated by setting the human among other kinds. In his descriptions of *behaviour* it is often difficult to discriminate human interpretation from physical structures. For example, discussing the secondary sexual characters of insects, he contrasts 'the pectinated and beautifully plumose antennae of the males of many species' with the meagreness of the females: 'the male has great pillared eyes, of which the female is entirely destitute' (2nd ed., p. 274). His children told him that his descriptions sounded like advertisements and here the males benefit from the enthusiasm of his language. And where he finds not physical difference but likeness between the sexes he comments, using the observations of colleagues, on contrasted behaviour:

> In one of the sand-wasps (*Ammophila*) the jaws in the two sexes are closely alike, but are used for widely different purposes: the males, as Professor Westwood observes, 'are exceedingly ardent, seizing their partners round the neck with their sickle-shaped jaws'; whilst the females use these organs for burrowing in sand-banks and making their nests. (p. 275)

—a striking example of separate spheres among sand-wasps.

Darwin clearly felt some little scepticism himself since he added a footnote stating, 'Mr. Walsh, who called my attention to the double use of the jaws, says that he has repeatedly observed this fact.' And fact it may be, since we cannot just wish away such structural and performative differences between sexes within particular species, even as we note the gendered interpretation being offered. Darwin's fundamental insistence in all his arguments on the similitudes between the human and other kinds inclines him to accept the fixed differences, for example in sand-wasps, as a model for human capabilities, rather than as the outcome of human behaviours in current social conditions.[6]

It is thus a relief to read at the end of the current volume the spirited and cogent challenge to Darwin's views from the Bostonian Caroline Kennard:

> In reply to your argument that "women must become as regular 'bread-winners" as are men"; have they not been and are they not largely bread-winners; though unrecognized generally as such? …
> The family must be *right*eously maintained Let the 'environment' of

---

6 These two paragraphs derive from my essay 'Late Darwin and the Problem of the Human' published by invitation on-line by the National Humanities Center as part of their project 'On the Human'. For the full essay and some responses see http://nationalhumanitiescenter. org/on-the-human/2010/06/late-darwin-and-the-problem-of-the-human/.

women be similar to that of men and with his opportunities, before she be fairly judged, intellectually his inferior, please.— (227)

Socially and intellectually, Darwin respected and delighted in the women of his acquaintance and drew on their specific skills and knowledge for his research, but he failed to observe in this one field the pressures of environment that were elsewhere fundamental to his arguments.

Preface

In *Descent of man* 2: 327, Darwin wrote:

> The chief distinction in the intellectual powers of the two sexes is
> shewn by man attaining to a higher eminence, in whatever he takes
> up, than woman can attain—whether requiring deep thought, rea-
> son, or imagination, or merely the use of the senses and hands. ... We
> may also infer ... that if men are capable of decided eminence over
> women in many subjects, the average standard of mental power in
> man must be above that of woman.

It was a surprising thing to write at a time when there was already much
discussion of the social disadvantages faced by women; their lack of education, their exclusion from the professions and politics, their legal disa-
bilities. Darwin's own beloved Jane Austen had pointed out, through her
heroine Anne Elliot in *Persuasion*, 'Men have had every advantage of us
in telling their own story. Education has been theirs in so much higher a
degree; the pen has been in their hands. I will not allow books to prove any
thing.' How could Darwin be unaware of the social bias that doomed most
women to underachievement, and the bias of perception that caused even
high achievers to be considered second rate compared with men?

Darwin knew of plenty of talented women through his correspondence
and in his daily life. There were women scientists who corresponded with
Darwin and sometimes even made a living of sorts in science. Darwin's letters also bring to light the participation of women in science in less public
ways, as editors, observers, collectors, supporters, and popularisers. This
activity is not always very evident in published works of Victorian science.
This book seeks to throw light on the lives of the women around Darwin:
what they were doing in science and other fields, and what kind of conversations they were having about women's rights and women's education. It
is a compilation of selected letters from the *Correspondence of Charles Darwin*,
the 30-volume edition of all known letters to and from Charles Darwin (the
Correspondence), publication of which is expected to be completed in 2022,
and from the collection of Darwin family letters in the Darwin Archive
at Cambridge University Library (CUL), most of which are unpublished.

Women made up about five per cent of Darwin's correspondents, and
letters to and from them about five per cent of Darwin's total correspondence. As is the case with his correspondence with men, not all of the letters

are about science. This book therefore aims to pull together various strands in order to highlight the contribution of women, who are often assumed, with a handful of exceptions, not to have been active in Victorian science, and examine their complex relationship to the developing institutions of science.

The Darwin Archive at CUL contains a great deal of correspondence between members of the Darwin family other than Darwin himself, and between members of the family and their own friends. In the Darwin Project, we refer to these as the family letters, to distinguish them from the letters published in the *Correspondence*. The distinction between the family letters and Darwin's letters (in the *Correspondence*) is loose. Letters written by members of the family to Darwin himself, or vice versa, are of course published in the *Correspondence*, as are letters not to Darwin that clearly contain information intended for him ('Tell Uncle Charles …'), and letters written on his behalf by other family members. Darwin probably read or had read to him most of the letters Emma received from family members as a matter of course, but only the ones he explicitly comments on in a reply are published in the main edition.

The family letters throw great light on the lives of the women in Darwin's life, and contain some surprises. Women writing to women, for instance, are often a good deal more forthright and unsentimental than women writing to men. Also, the family letters reveal interests and pursuits that are rarely mentioned in the more familiar sources. Views on education, feminism, politics, charity, religion, and marriage, for example, are discussed in the family letters far more than in Darwin's own correspondence.

The letters, both Darwin's letters and the family letters, help us to understand something not at all evident in Darwin's published work: the social substructure of science and the extent to which it was underpinned by unacknowledged female activity, and, sometimes, by unacknowledged male activity. Female obscurity was both a consequence of the social status of most of the women concerned (high rank made a great difference to women's willingness to accept publicity), and of their own choices. Women's observations are often recorded in Darwin's books as coming from 'a lady' or 'a friend', whereas men are likely to be named. This is not a sign of disdain on Darwin's part. Some letters survive in which Darwin asks his informants how they wish to be cited, and it's likely he always asked, if possible. Published authors, whether male or female, would naturally be cited by name, but private persons were allowed to choose.

Scientists depended on social gatherings to cement professional relationships, and these were often organised by women. Mary Lyell and Ellen Lubbock were prominent hosts. Emma Darwin played only a small role in this arena; her main concern was to see that Charles avoided stress as much as possible, and he found social gatherings stressful. However, she sometimes collaborated with the wives of Darwin's closest colleagues: Henrietta Huxley, Ellen Lubbock, and Frances Hooker: often, as we see in the chapter 'Scientific wives and allies', this activity had to be obscured after the event. 'My name must not appear', must have been a frequent plea of female activists. The role of women as go-betweens was complex. When Lord Derby received a petition to preserve the land around Niagara Falls, it had passed through the hands of Sara Darwin, William Darwin, Charles Darwin, Emma Darwin, and Lady Derby. If Emma and Lady Derby could not provide a conduit to Lord Derby, William told Charles, it would be no good, and the petition had better not be sent, even though Charles had already corresponded directly with Lord Derby. On the other hand, when a woman's name appeared on the list of contributors to a financial appeal on behalf of Thomas Huxley, Darwin's friend John Tyndall was dismayed: it looked as if there had been an 'effort'.

Darwin worked and received scientific visitors at home: the work of keeping the household running smoothly was masterminded by Emma, with the help of her staff and (as they grew older) her children. It could be because she did her work so well that we hear so little of it in Darwin's letters. (A series of letters from Darwin's friend Joseph Dalton Hooker in the chapter 'Servants and governesses' shows how overwhelming a preoccupation housekeeping could become if a wife was too ill to take care of it.) Here, the family letters are crucial to understanding what was going on behind the scenes: for instance, the social alarm that ensued when it looked as if two ornithologists, one Russian, one American, were about to visit at the same time as Lady Derby. (Lady Derby was put off.)

Looking at the letters between women sheds light not only on their lives but on Darwin's own life. When she married Darwin, Emma not only took a share of his responsibilities for keeping in touch with friends and family, but became an avenue for messages from people who didn't want to bother Darwin himself. If Darwin had not married, his correspondence would probably have been larger and more varied in tone. The letters in the *Correspondence* are on the whole scientific because Emma was writing most of the chatty, newsy, keeping-in-touch letters, the ones about birth, death, marriage, travel, clothes, servants, and all the rest of it. Darwin would have read or had read to him many letters that weren't addressed to him, and no doubt knew the content of many that weren't signed by him. Long letters were often circulated around the family to save the writer the bother of writing the same things many times. (Abbreviations and symbols are also common in the family letters: probably the Victorians would have loved social media and text messaging.)

It wasn't that Darwin couldn't write a newsy letter; he kept in touch with his cousin William Darwin Fox and his old shipmate Bartholomew James Sulivan by this means, and the chapter 'Friends' shows that Darwin could and did cultivate female friendships: but he didn't do very much of this sort of letter-writing, because like most married men whose wives and children were close at hand most of the time, he didn't have to.

Regular letter-writing, for no particular purpose other than keeping in touch, was somewhat more of a female than a male art-form. Jane Austen in *Mansfield Park* is humorous on the subject of women's hoarding of titbits of news to spread over pages of a letter, while men's letters are, stereotypically, brief and to the point, if written at all (the Darwins and Wedgwoods were steeped in Jane Austen). When Darwin was on the *Beagle*, his sisters promised to write at least one letter a month, and on the whole managed to keep it up. His brother Erasmus, however, had to have his arm twisted to extract a letter: Catherine Darwin joked that brothers never could write to each other (*Correspondence* vol. 1, letter from Catherine Darwin, 26–7 April [1832]). Charles found that the best way to get a letter from Erasmus was to give him 'commissions' (things to do) so that he would have to write about how he got on about them. Erasmus's surviving letters often contain only a sentence or two. Darwin was much more chatty, but still, he usually only wrote when he had something to say, or had been prompted by receiving a letter himself. Consequently, when we read Darwin's life through his letters, we get a rather partial view. Darwin had no notion of diarising his inner life through letters, and except when he knew his letters would be published, as with the ones he wrote on the *Beagle* voyage, he didn't diarise his external life in much detail either.

This volume is different from previous selections of letters published by the Darwin Correspondence Project as a supplement to the main edition in that it is based on a theme, women, and has thematic rather than chronological chapters. A different point of focus could have been chosen and probably will be chosen by future scholars: the correspondence could be used from the point of view of historians of class, empire, or childhood; or it could be looked at from the point of view of nationality (a Calendar of the German correspondence has already been published). Because of the thematic arrangement of the chapters, it has been necessary to excerpt letters and provide more explanatory text than appeared in previous selections. A typical letter from Emma to one of the children, for example, would normally range over a number of subjects: the weather, what has happened at Down, what news has come from relations, advice on some problem raised by the addressee, comments on current affairs, plans for the future, and so on. In addition, more biographical information about the correspondents and information from other sources, including their own publications, has been included.

The themes that formed the chapter headings for this book were principally derived from an analysis of the subjects on which women wrote in the letters in the main edition of the *Correspondence*. About half of women's correspondence turned out to be what might be called friends-and-family letters: chatty, newsy, keeping-in-touch or staying-organised letters between Darwin and female relatives and friends.

Of the remaining letters, the next biggest group was from observers in the widest possible sense, from women working at a high level in science to mothers noting at Darwin's request at what age their baby first cried real tears. The majority of observations came from botanists. Botany was the female science par excellence in Victorian times, and the chapter 'Observing plants' could easily have been made three times the size of the others, and still only included a handful of serious botanists, without even starting on the casual or unskilled observers. Other observations were on animals, both domestic ('Companion animals') and non-domestic ('Insects and angels'). Finally, there were observations on humans: mothers who had been asked to watch for their babies' first tears; women responding to Darwin's questionnaire about expression of the emotions or to his book, *Expression of the emotions*; and letters about supernumerary digits. There were also letters from writers; letters from editors (principally Darwin's own editor-in-chief, his daughter Henrietta); a handful of letters on religion; and a small number on women's rights ('Ascent of woman').

Some chapters were more challenging, such as the one on travel. The notion of the Victorian women traveller, marching into the jungle armoured with her veil and a 'good stout skirt' is very familiar to modern readers, but very few of that sort of traveller wrote to Darwin. Henrietta, Darwin's daughter, was an indefatigable traveller in Europe, where she tramped around, usually with a relative as a companion, and wrote exhaustive and sometimes exhausting accounts to her parents at home. This sort of travel was surprisingly common for Victorian invalids and hardly merited letters to the Geographical Society, but it was travel, none the less. Henrietta's account of a train wreck in which she was involved is included in this chapter. Lady Florence Dixie was a Victorian traveller more in the classic mould, a big-game hunter in search of solitude: she wrote to Darwin about her observations of animals in South America and to send him a copy of her book, written in the interval between her expedition to Patagonia and her departure for South Africa to be a war correspondent. Marianne North, painter and traveller, visited Darwin at Down as his request: only one short letter from Darwin to her survives, through the medium of her own autobiography.

The chapter on servants emerged not from reading letters to or from Darwin but from reading the family letters, where they play a much larger role. Choosing, training, keeping, and sometimes maintaining lasting relationships with servants was a major preoccupation of housewives. One harrowing episode involving a fierce governess is mentioned only in the family letters, even though Darwin himself did have to intervene. Having to sack a servant gave even the strong-minded Henrietta the jitters: it was a relief to have a man in the house at the time, according to her, even though the man wasn't going to do the actual sacking. Once the theme had been established, on looking again at the main edition, a series of letters from J. D. Hooker to Darwin detailing the trouble in his household when his wife was ill and unable to supervise the female staff had added force. This theme is doubly important since Darwin thought that women would only equal men in intelligence when they became breadwinners. Reading Darwin's own correspondence, it's easy to forget that he was surrounded by women earning their own living, at various different levels of society. (Darwin himself lived for much of his life mostly on unearned income. Nevertheless, when his cousin Francis Galton sent him a questionnaire about his character, it was his acumen with money that he was most proud of: apart from that, he didn't claim that there was anything special about his intellect. Also he was extremely proud of the amount of money his books made.)

Many of Darwin's female correspondents were prominent feminist campaigners and radicals, and based on their biographies, it seemed inevitable that there would be a chapter on Victorian feminist thought. In fact, this chapter ('Ascent of woman') was not straightforward since hardly any of these women wrote to Darwin about their thoughts or their campaigns. Lydia Becker, author of an article 'Women in science' in the *Contemporary Review*, and a leader of the women's suffrage campaign, wrote to Darwin about botany. Eliza Meteyard, radical and feminist, wrote to Darwin about his Wedgwood relations, and asked for his support in her petition for a civil-list pension. The only surviving direct challenge Darwin received on his views about women's intelligence was from Caroline Kennard, an American. She wrote first to make sure she'd understood him correctly. When he replied, confirming that he thought women would not equal men in intelligence until they were breadwinners, and adding that their becoming breadwinners would tend to detract from the happiness of the home, she answered briskly: women already were breadwinners, albeit severely handicapped by lack of educational and professional opportunities. And was work any the less work because it was unpaid?

Despite the rarity of letters showing women's direct engagement with Darwin on this issue, the family letters, some of which are included in this chapter, reveal Darwin's relations' awareness of feminism. Henrietta was on the edge of liberal and feminist movements—she met Josephine Butler—but was not necessarily a convert. Elizabeth, Darwin's other surviving daughter, who is often underestimated, went to lectures at London University. Some of Darwin's female relations and their friends seemed bent on learning difficult subjects and taking examinations. Amy, Darwin's daughter-in-law, wanted to learn mathematics as a route to physics. Emma, under the influence of her niece Snow Wedgwood, thought it would be a good idea if women had the vote, largely because she thought this would influence MPs to take a more favourable view of legislation against cruelty to animals. Elinor Dicey and her husband, family friends of the Darwins, helped set up Newnham College, a women's college in Cambridge. Snow Wedgwood, Darwin's niece, taught in Hitchin at a precursor of Girton College, another Cambridge women's college. Erasmus Darwin was chairman of the council of Bedford College for Women in London; Emma's sister-in-law Frances Wedgwood was also on the council. Darwin was surrounded by thought, talk, and action on 'the woman question', as Victorians called it.

Darwin was not overtly opposed to women's higher education. Towards the end of his life, events seemed to catch up with him when women's influence in the anti-vivisection movement became clear. He thought women underestimated the medical benefits of vivisection, and attributed this to their lack of education in physiology. He became a supporter of physiological education for women. In 1881, women were given the right to take examinations at Cambridge University. Darwin commented in a letter to his son George: 'You will have heard of the triumph of the Ladies at Cambridge. The majority was so enormous that many men on both sides did not think it worth voting. The minority was received with jeers. Horace [*Darwin*] was sent to the Lady's College to communicate the success & was received with enthusiasm.' (DAR 210.1: 103.)

When Darwin wrote on women's capabilities in *Descent*, he couldn't have been unaware of the problems of bias and social disadvantage. However, he kept his argument strictly biological, and he relied on two factors he himself wasn't entirely sure about: the inheritance of acquired characteristics, and inheritance limited by sex. His theory was that men had long undergone more severe testing than women as adults in competition for wives, and that the qualities they acquired as a result—energy and perseverance—were passed on to their sons but not their daughters (the principle being that qualities that manifest later in life tend to be limited to one sex: as, for instance, colourful plumage in the male peacock). This accounts for his notion that for women to equal men in intelligence, they would not just have to be educated as young adults; they would have to pass on the effects of that education to their daughters and repeat the process for many generations.

Even given Darwin's beliefs, it's not a good argument. Intelligence does not manifest only in adulthood, like the peacock's tail. Many women must have doubted, like Charlotte Papé, that the men around them were really cleverer, on average, than the women, or, like Caroline Kennard, that men were really making more of an effort. Still, Darwin's account did at least suggest that change was possible. Even though female inferiority might be written into their biology, their biology could change. Nor was Darwin particularly dogmatic about it: in his reply to Kennard, he reiterated that his beliefs were based on laws of inheritance that he only hoped he understood correctly. Savages, he thought, were more equal; maybe civilised people could be too.

*****

A few notes on the text: in order to keep the text as readable as possible, footnotes have been avoided; instead there are a minimal number of short clarifying notes in the letter texts themselves. These are in square brackets and italics. Omissions are marked by ellipses (...). For full texts of letters to and from Darwin, with explanatory footnotes, and for some third-party letters, see *The correspondence of Charles Darwin* or www.darwinproject.ac.uk. (Some letters may not yet be available.) References to works in the bibliography are generally in author–date form (e.g. Becker 1864), but some reference works and Darwin's own publications are referred to by short titles (e.g. *ODNB*, *Descent*, 'Climbing plants'). Again, for readability, these have been kept to a minimum: there are some publications in the bibliography that are not formally referenced in the chapters, but that may be useful for further reading. The spelling and punctuation of the original letters have not been altered. Underlined words in the original texts are reproduced in italics, and double underlined words in bold. Printed addresses, from headed notepaper, are also in italics.

Some letters could have appeared in more than one chapter; for instance, a letter from a woman who had lived in India about the people and animals she encountered there could have appeared in three different chapters. (In fact, it's in 'Travellers'.) The decision of where to include letters is fairly arbitrary, being influenced mostly by the need to keep the chapters a reasonable length. There are cross-references to related material in another chapter that the reader might like to see. Likewise, when a woman's letters appear in more than one chapter, there are cross-references to highlight related material.

Brief biographical details, if they could be discovered, for all correspondents and some of the other persons mentioned are included in the Biographical notes. Full details of the date, provenance, and previous publication of each letter of which substantial portions are reproduced are given in the List of letters; otherwise brief details are given in the text.

Acknowledgments

This publication by the Darwin Correspondence Project rests on the work of members of the Project both past and present, in finding, transcribing, proofreading, and researching the texts included here. Their names, and the names of the many librarians, archivists, and other experts who have assisted the Project are included in the acknowledgments in the volumes of the main series of the *Correspondence of Charles Darwin*.

Particular thanks for their help in making this book a reality are due to: Philippa Hardman, for assistance in several drafts of the original proposal and for her invaluable knowledge of the field; Jim Secord, for suggesting a thematic arrangement, for commenting on the first draft, and for his constant encouragment; Charissa Varma for contributing the chapter on children, for many helpful suggestions about the other chapters, for tireless work reading and transcribing letters and suggesting background reading; Paul White for the chapter on religion, only one of the subjects on which he has become an expert; Rosemary Clarkson for writing and keeping track of the many permissions letters, for proofreading, and for her skills as an Ancestry.com wrangler; Elizabeth Smith for proofreading letter texts and keeping a careful eye on the typesetting; Andrew Corrigan for finding and preparing images and acting as typesetting consultant; Gillian Beer for writing the introduction; and Margot Levy for preparing the index.

Thanks are also due to our editors at Cambridge University Press, Linda Bree and Anna Bond.

Finally we would like to thank the owners of letters and the owners of copyright in the letters for their kind permission to publish them here. Every effort has been made to trace holders of copyright in letters written by persons other than Darwin where copyright permission is required for publication. The Darwin Correspondence Project is extremely grateful to families and estates of letter authors for permission to include their works in this publication, and particularly to William Huxley Darwin and other descendants of the Darwin family for permission to publish the texts of letters written by the Darwins. Details of repositories that supplied copies of letters are in the List of letters and provenances at the end of the book.

Symbols, abbreviations, and conventions

CUL	Cambridge University Library
DAR	Darwin Archive, Cambridge University Library
⟨ ⟩	Damaged text
[*text*]	Editorial note
|	New line in original text
text	In letters: underlining in original text/printed address
text	Double underlining in original text

Full details of letters featured are in the List of letters and provenances, pp. 229–40.

1   Friends

Darwin loved female company. As a boy in Shropshire, he spent time not only with his sisters and Wedgwood cousins, but with the Owen girls at Woodhouse. Later in life, Emma Darwin was entertained to see him flirting prettily, as she put it, with female visitors. He was on cordial terms with the ladies he met while he was undergoing hydropathic treatment, and Ellen Lubbock and Henrietta Huxley sent him teasing, funny letters. The formidable Lady Derby kept up an intermittent friendship with him in a series of visits and characteristically brief letters. As a old man, Darwin made an effort to reconnect with the Owen girls, sending a copy of his book on expression of the emotions to the elderly Sarah Haliburton, as she had become.

The first letter is to Darwin from an elderly friend of his family, Mary Congreve. At the time, in 1821, she would have been 75; Darwin was 12. Little is known of Mary. Her brother William, comptroller of the Royal Laboratory at Woolwich, where ammunition was manufactured, became a baronet, and his son, William, the second baronet, became famous as a rocket designer. It's tempting to suggest that the Congreve family might have fostered Darwin's youthful interest in chemistry.

My dear M͏ͬ Charles
I find I have only just time to thank you for your entertaining letter, as if I take time to write what I intended I shall not be able to get it franked & I'm sure it will not be worth the postage, I should have liked to have seen the good Gentleman *Grin* that you mention   there is no doubt but those that were out of the Scrape were much amused, I assure you I wish'd much you had been of our party on thursday night at the play, I think you would have been highly entertained both with the Coronation, and the entertainment of Monsieur Tonson [*a farce by W. T. Moncrieff*], I never laugh'd so much at a play I think, I dare say you have been much amused with M͏ͬ Alexander [*a ventriloquist*] & I hope I shall hear some specimenes of his art from you when I return, as I dare say it is practiced in School Lane, so god bless you as I am obliged to conclude this ever believe me | Yours truly M Congreve ...
I think you will not be able with all your Greek knowledge to read this precious Scrawl

Darwin was a boarder at Shrewsbury School, close enough to home to see his family regularly. In 1825, at the age of 16, he went to Edinburgh University with his brother, Erasmus, to study medicine, but soon decided it was not for him. He then spent three years at Cambridge University, with the intention of later becoming a clergyman. In 1831, he was invited to join HMS *Beagle* as companion to the captain on a surveying voyage to South America and circumnavigation of the globe. When Darwin departed on the *Beagle* voyage, Fanny Owen of Woodhouse wrote her farewells in a letter of 26 September 1831. She had kept up a long, jokey correspondence with Darwin whenever they were apart, alluding to the games they played as children.

> 2. Northernhay Place, Exeter
> Monday
>
> My dear Charles,
>   I have this evening heard from Caroline that you leave home the end of this week—and that you wish to have a *good bye* from me before you go. I had not the **least idea** you were to go so soon, for they told me it was the end of October you sailed, so I **hoped** and fully expected I should have been at home in time to see you— I **cannot** *tell you* how *disappointed* & *vexed* I am that that cannot be. Little did I think the last time I saw you at the poor old Forest [*Woodhouse*], that it would be **so long** before we should meet again!! This horrid Devonshire—fool that I was to come here— I shall just get home when you are gone I dare say— My dear Charles I do hope you will enjoy yourself & be the happiest of the happy, I would give any thing to see you once more before you go, for it does make me melancholy to think the time you are to be away—& Heaven knows what may have become of all of us by this time two years. at all events we **must** be grown **old** & steady— the pleasant days, and fun we have had at the Forest can never come over again— how I wish I was there this week to have one *last chat* with you   I cannot bear to think you are really going *clear* away, without my saying one *good bye*!!
>   But I must drop this subject for I find I am getting prosy & melancholy & that wont do— They tell me you were at Plymouth about 10 days ago & so was I, how **very very** unlucky we never met, do you go there again? if you should perhaps you may pass through Exeter— I shall leave it on the 6th with the Hunts— I believe not come home direct but go with them to pay some visits— if possible I shall shirk and get the Gov— [*governor, i.e. father*] to meet me at Leamington or Birmingham for I think it will be awful *flat work*, dowagering about with the Hunts to unknown parts— I am sure I have been dull enough all this summer— hope I have expiated all my sins for a severe Penance I have had of it— I wont be *taken alive* again in that way when once I get home— *Home sweet home* you should hear me sing now— I assure you I do it **feelingly** —it would melt a heart of stone—or rather crack an **ear drum** of **Iron** to hear me—but here my powers

have no scope I can never give vent to my feelings as I feel inclined—
… did you throw yourself on the Governor's mercy, & confess your
creditors, or what have you done? What a capital way of escaping
ungentlemanlike Tailors &c— When you are *far from the Land* they may
whistle for their cash for what *you care*! Well, dont be surprised if you
hear I have *taken Ship* too and fled my duns— that **joyful** season
Xmas is fast approaching— my heart sinks when I think of it—but
there's nothing like putting a good *face* on it— I shall do so as long as
I can— Pray write to me one last Farewell my dear Charles & tell me
all your plans & prospects—where you are to go to—& all about it?
And tell me too if I shall look out for a nice little Wife for the *Parsonage*
by the time you return. tell me what you require and I will look about
and get one in *my eye* by the time you want her—a proper *knowledge*
of the *Beetle tribe* of course you require— bye the bye has *your faithless*
Charlotte Salway bee⟨n⟩ twined off yet—I have heard nothing of her
As for all your Sisters I think they are gone crazy or *sulky* or sleepy
or somethi⟨ng⟩ for not one line have I had from any of them these
two months—they treat me with the most marked contempt.— I was
much amused at Plymouth there is so much worth seeing— Mount
Edgecombe I dare say you saw—it is a beautiful Place.— I went on
board the Adelaide and all over it—so can fancy you in your little
Cabin—and I assure you you will not be forgotten, I shall often long
to have you to laugh with and *scold* out of the Painting room— I
wish I had made your Pincushions they might have been useful—and
occasionally in taking out an *instrument of death for a Beetle* you would
have called to mind the Manufacturer of the *useful article*—but it cant
be helped now— this letter is *most prosy*, & duller than letter ever was
before—but I cant help it you must take the *will for the deed* — write to
me 2 Northernhay Place= I must now conclude—can only add—I
most sincerely wish you every amusement & happiness possible—but
only wish most heartily you were not going quite so soon that we
might have one *more talk & laugh* first—but it is *not* to be— so good
bye my dear Charles
    Believe me always yours most sincerely and *affecty* | F O—
    Burn this before *you sail for pitys sake* —

By the time Darwin returned, in 1836, Fanny was married and had a
daughter. She wrote more soberly to thank Darwin for a gift of flowers on
14 January 1837.

My dear Charles,
    I am ashamed to think how ungrateful I must have appeared to
you—for I believe it is more than a month since I received your beautiful present of Flowers & they have remained quite unnoticed by a
line of thanks.— pray forgive me I have indeed been more or less so
unwell since I received them that I have not been able to write or do

any thing else— accept now my best thanks, I was *very much* pleased by your kind recollection of me— the Flowers are the prettiest things I ever saw, much too good to wear I think & I mean to do justice to them in a *glass case*—

—I think you have used your friends very shabbily in taking flight so soon again. I had no idea you were going away for the whole winter— I hope when you have any *precious* time to throw away you will find your way to Chirk Castle— where I assure you we shall both be delighted to see you—

ever dear Charles y^{rs} most truly | F Myddelton Biddulph

Chirk Castle Jan^{ry}. 14th. 1837.

In 1838, Darwin married his cousin Emma, and after four years in London moved to Down in Kent. During this time there is little surviving correspondence with women other than members of his own family: Darwin was frequently ill and when he was not was busy writing and studying, establishing himself as a respected man of science. In 1849, he began to visit hydropathic establishments in search of a cure for his ill health. At two of these establishments, Moor Park in Surrey and Ilkley Wells in Yorkshire, he encountered Mary Butler, the sister of Richard Butler, the vicar of Trim in Ireland. Butler visited the Darwins at Down in 1860. Evidently she and her friend, the novelist Georgiana Craik, had discussed Darwin's theories with him. With the first surviving letter, written early in 1859, Darwin sent autographs, no doubt cut from the letters of his naturalist friends:

Down Bromley Kent
Feb. 20th

My dear Miss Butler

I send you some autographs with a list of the men, as you, perhaps, would not know who were who. You will now be well stocked with the autographs of *Naturals*.

I made myself very pleasant at home with ghost stories & other plumes borrowed from you.

I enjoyed my fortnight extremely at Moor Park, but if I were long exposed to the very pleasant temptation of sitting between Miss Craik & you, I wonder what I should not come to believe: Honeysuckles turning into oaks would be a mere trifle & new species springing up on every Railway embankment.

Will you tell Dr Lane that I found Etty [*Henrietta, Darwin's daughter*] looking as well & as fat as before her illness.

Pray give my kindest remembrances to all the very pleasant party at Moor Park & believe me with much respect | My dear Miss Butler | Yours Truly obliged | Charles Darwin

Please to tell Lady Drysdale that I reached the Station only 14 minutes before the Train started & I should like to know when she will ever have such a triumph as that.

Later in the same year Darwin wrote to find out whether he could expect to see her at Ilkley. He was finishing work on *Origin of species.*

Down, Bromley Kent
Sept. 11th

My dear Miss Butler

I wrote to Moor Park to enquire for your address, & was told that a letter addressed to you at Mr Tennant's would be forwarded, but that you were wandering about Scotland. This, I much fear, augurs badly for Ilkley.— My Book at last is so nearly finished that I can really & truly see that I shall be a free man at the end of this month. Our plans are rather undecided; but I incline strongly to go to Ilkley, but I fear, without I found it a very tempting place, that it is too late to take a house for my family; & in this case I should stop three or four weeks in the establishment, return home for a week or so, & then go to Moor Park for a few weeks, so as altogether to get a good dose of Hydropathy.

My object in troubling you with this note,—a trouble, which I hope & believe you will forgive—is to know whether there is any chance of your being at Ilkley in beginning of October. It would be rather terrible to go into the great place & not know a soul. But if you were there I should feel safe & home-like.— You see that all your former kindness makes me confident of receiving more kindness.

I hope that you are well & have had happy visits with your friends,

Pray believe me, my dear Miss Butler, with truth | Yours sincerely obliged | Charles Darwin

In December 1862, Butler wrote her last extant letter to Darwin, asking for assistance for another fellow patient, Mr Thom. Darwin sent him £20. This letter is doubly interesting since 'asking for money, or a job, for someone' is a small but significant theme in letters to Darwin from women. Darwin had grown a beard by this time, at Emma's suggestion, possibly to reduce the irritation caused by shaving and eczema. Darwin wrote to his son William in July 1862, 'Mamma says I am to wear a beard.'

Sudbrook Park | Petersham
Wednesday

My dear Mr Darwin

We were all relieved & made happy by Mrs Darwins account of you and I wish that I could go to you now and have the very great pleasure of being once more amongst you all, but Lady Drysdale some time since, made me promise not to leave Sudbrook till after

Christmas— We are a quiet sociable party here, & the absence of even one would make some difference in the arrangements of the house. I have an interest much at heart just now, which I fear you will not be able to assist—willing to do so I am certain you will be— You remember Mr Thom—who excited y^r. admiration by the several victories which he achieved over Brandy, Opium Tobacco—& himself!! he has been the steadiest of men ever since Clever, well educated, highly principled—modest!— For some years he has been nominally Sub Editor of the Home News (from which M^r. Robert Bell derives the revenue) but really the sole Manager of the Paper, which has an extensive circulation, & is said to be extremely well conducted—

The Sedentary life in a damp office in the City has so completely undermined his health that he is obliged to give up his employment, & has no prospect of meeting with a suitable one in this Country— so that as a last resource he is going to Queensland—at the age of 33—to spend the remainder of his days amongst Cows & Sheep (he scarcely knows one from the other) in a strange Country—where he has not even a friend; I feel for him deeply

There are many persons to whom the services of such a man would be valuable—for his abilities are excellent—& he has the highest testimonials as to character, whilst both his appearance & manner are prepossessing— It has struck me as *just possible* that you may know of some place to fit him. He has been trying to get into the Constabulary but has not the proper interest to give any hope of success— he tried for the Secretaryship to an Hospital & found that there were Six hundred Candidates!!— An Inspector of Schools he once thought of, but of that there is no chance, from the Government Interest requisite— Could you speak a good word for him in some influential quarter? he would do you no discredit I believe in any way, for he is really a superior & meritorious man?

We have felt a good deal of anxiety about M^r. Smyth—who was thrown *penniless* upon the kind family here—but M^r. Tennant of Glasgow has given him an appointment in Trinidad where he has a prospect of becoming a Planter & doing well—he sailed for the West Indies a fortnight ago.— My dear M^r. Darwin I will not excuse myself for writing all this to you, I scarcely believe in the possibility of your having it in your power to befriend poor Mr Thom, but at all events I am assured of your most kind sympathy—

My best love to M^rs. Darwin. I dont like the idea of your long beard. M^r. Davenport who is here—wears one from the same cause, but he has benefited wonderfully from the frequent use of the Turkish Bath—& is beginning to look perfectly handsome—

Always Sincerely & affectionately Yours | Mary Butler

The Darwin and Huxley families became close not only because of scientific sympathies between Charles Darwin and Thomas Huxley but because they both had many children, and the Darwins, who were older and better established, gave the Huxleys a good deal of practical and emotional support, on occasion transferring the whole family from London to Down for a rest. Henrietta Huxley, Thomas's wife, liked to tease Darwin with literature.

Dear M: Darwin

Hal has just brought me your note containing your slyly disparaging remarks on my beloved Tennyson—& quoting "as a gem"

'And he meant, he said | he meant, | Perhaps he meant, or partly | meant you well.'

In the first place it was very mean of you to give the lines without the context shockingly Owenlike [*an allusion to Darwin's adversary, the anatomist Richard Owen*]

Secondly. The lines only convince me more than ever that Tennyson is quite master of his situation. Could you better render In words, the desire in the wife's mind to do justice, to—her enemy I suppose for I have not read "Sea Dreams", together with the conflicting feeling which yet possessed her of his insincerity? I am very pleased that Tennyson accredits the feminine mind with such a strong sense of justice.

I now refer to the book— I am grieved to find that a philosopher of your repute—should have damaged your reputation for accuracy so greatly as to tell me that the quotation was from "Enoch Arden" whereas it was from "Sea Dreams"— If the "facts?!" in the Origin of Species are of this sort—I agree with the Bishop of Oxford— [*Samuel Wilberforce had criticised the Origin of species at the British Association for the Advancement of Science meeting in 1860.*]

Yours too sincerely | Henrietta Huxley

love to your dear wife & ask her for a screed.

New Year's Day | 1865.

In 1872, the Huxleys moved house.

at Miss Woodington's | The Common | Sevenoaks
Oct. 16th

My dear M: Huxley

Every man has a right to give a friend a marriage present; & going into a new house is nearly as serious & dangerous an affair as marriage.— Therefore I have a full right to enjoy the pleasure of making you a marriage present. I defy your husband, with all his sharpness, to pick a hole in this logic. But here comes my difficulty: I want to give something useful & not poetical, & I thought of asking to be allowed to furnish your dining room; but then I know not what furniture you already have. Now will you not allow me to treat you, as I have treated some of my near relations (& I am sure that I feel like a near relation to you all) & ask you to buy something with the enclosed for your self.—

Do grant me this favour.— I was very sorry to hear so poor an account of your husband's state, both for my own sake, & you must know what admiration & affection I feel for him, & for the sake of the whole world.— I hope that he may soon improve, & there is at least one comfort in indigestion, with all its miseries, that there is always a good chance of a prompt cure.—

Pray believe me, my dear | Mrs Huxley.— | Yours affectionately | Charles Darwin

Mary Catherine Stanley, Lady Derby, was the daughter of George Sackville-West, Earl De La Warr; she married James Gascoyne-Cecil, the marquess of Salisbury, and after his death she married Edward Henry Stanley, the earl of Derby. Her *ODNB* entry describes her as a politician manqué, and speculates that her childhood friendship with the duke of Wellington might have been the source of her fascination with 'politics, diplomacy, and war, and her preference for male conversation'. Her second marriage brought her to Holwood House, in Keston, Kent, not far from Darwin's house at Down. Her letters to Darwin are notably brisk and brief. They begin with a shared interest in the writings of the psychic investigator William Crookes, on 16 November 1871:

> Holwood | Beckenham
> Thursday Evg
>
> Dear Mr Darwin
> I could not lose a moment on my return home—& read the article most eagerly. You will be obliged to believe that Mr Crookes has "a craze". It staggers **me** a good deal.—but I know that my imagination is apt to overpower my judgment!—
> I wish I had seen you after you had read the article. I sh^d have liked so much to hear what effect it produced on you!—
> Y^rs very sincerely | M C Derby

> Down | Beckenham | Kent
> Saturday
>
> Dear Lady Derby
> If you had called here after I had read the article you would have found a much perplexed man. I cannot disbelieve Mr Crookes' statement, nor can I believe in his result. It has removed some of my difficulty that the supposed power [*altering the weight of objects remotely*] is not an anomaly, but is common in a lesser degree to various persons. It is also a consolation to reflect that gravity acts at any distance, in some wholly unknown manner, & so may nerve force. Nothing is so difficult to decide as where to draw a just line between scepticism & credulity.

It was a very long time before scientific men would believe in the fall of aerolites [*meteorites*]; & this was chiefly owing to so much bad evidence, as in the present case, being mixed up with the good.

All sorts of objects were said to have been seen falling from the sky—

I very much hope that a number of men, such as Professor Stokes will be induced to witness Mr Crookes' experiments.

Pray believe me | your Ladyship's | truly obliged | Charles Darwin

23. St James's Square. | *S.W.*
June 4/72

Dear Mr. Darwin

Sackville [*Sackville Cecil, Lady Derby's son*] would be extremely pleased to be allowed to be present with Mr Galton at a Séance of Mr Crookes'.—tho' he doubts being able to form any opinion without going thoroughly into the Evidence, & this,—with the work he has in hand would not be possible.

But the truth is I am very eager Sackville should be at one of Mr Crookes' séances, & if you think it likely Mr C. w^d allow him to go with Mr Galton—w^d it be asking too much of you to try to arrange it? Sackville is very sceptical on the point but very curious— I am all ready to hear of a new force & very curious indeed.

…

Believe me | Y^{rs} very sincerely | M C Derby

As an occasional neighbour of Darwin's, Lady Derby took an interest in the district, but was rarely able to visit. In this letter, a visit from her has almost coincided with a visit from two ornithologists, one Russian, one American.

Fairhill, | *Tunbridge.*
Sept 14/75

Dear Mr Darwin

It was very good of you to write to me yest^y & I thank you much for telling me such exact truth. I was very much disappointed not to go to Down, but sh^d have been in despair had I found myself arriving at an inconvenient moment. I must now defer my visit till November, for we go to the North early next week.

I went on to Keston to see Mr Carlyle; the country air has done him great good & I want him to linger on at Keston till the fine weather leaves us. [*Lord Derby had put Keston Lodge at Thomas Carlyle's disposal for the summer of 1875.*] I suspect he is getting rather dull, & is half sorry to have been so unsociable to his neighbours on his first arrival!

I was in the New Forest the other day & saw some birch trees with bark exactly like that of the birch in Holwood which I remember hearing you speak of.

Believe me | dear Mr Darwin | Yrs very sincerely | M C Derby

I hope Mrs Darwin's headache has passed away

In 1875 she wrote to thank Darwin, probably for praising Lord Derby's inaugural address as rector of the University of Edinburgh.

Knowsley, | *Prescot.*

22 Dec^r/75

Dear Mr Darwin

Though you tell me not to answer your most kind note I cannot help disobeying you: Your warm & genuine expressions of approval have given Ld Derby more pleasure than any other compliment he has received, & you must forgive me for saying so.— We made two short visits to Keston last month, but I was never able to find time to get as far as Down.

We are more & more pleased with Keston each time we go there.

Will you remember me kindly to Mrs Darwin & believe me | Y^rs very sincerely | M C Derby

Knowsley, | *Prescot.*

19 Sept/77

Dear Mr Darwin

Count Schouvaloff [*Peter Andreivich Shuvàlov, Russian ambassador to London*] has been asserting today that your works are still prohibited in Russia. I told him your story as you told it to me, but he thinks I have made a mistake. If you would not mind dictating a letter to me stating what you believe to be true, I sh^d be much interested to be able to tell him that *he* was mistaken.

Still if you prefer to let the matter alone take no notice of my request.

Yrs very sincerely | M C Derby

23. St. James's Square. | *S.W.*

May 24/78

Dear Mr Darwin

My brother who has just returned from S. America has brought from the River Plate the accompanying fragment of bone from a fish's head called *Corbin*; he is very anxious to know if it ever came under your notice. There are two of these bony substances in the head of every fish. Fibrous threads diverge from the rough part in the interior—as if this substance were the covering of the brain! if one can venture to speak of the brain of a fish.

Forgive me for troubling you & for daring to suppose I am mentioning any thing that can be new to you.—

Believe me | Yrs Very sincerely | M C Derby

The following letter, from 1879, shows a message from Darwin to Lord Derby being transmitted in a letter from Emma to Lady Derby.

Down, | Beckenham, Kent. | Railway Station | Orpington. S.E.R.
Tuesday | Nov 12

My dear Lady Derby

My eldest son has received the accompanying papers from Mr Olmstead (so distinguished for his services in the American war)

He is very anxious to obtain some influential signatures to the petitions & Mr Darwin sends it to you in hopes that Lord Derby may be inclined to give his— I enclose an envelope to return it.

My husband sends by this post a short notice of his grandfather Dr E. Darwin which he has just published. He would be much pleased if it interested you in any degree—

Believe me | my dear Lady Derby | very truly yours | Emma Darwin

The petition, which originated with Frederick Law Olmsted and Charles Eliot Norton, had to do with buying up property around Niagara Falls so as to provide better public access and preserve the site for future generations. Darwin had received it from his son William, who probably had it from his American wife, Sara, Norton's sister-in-law. William wrote to Darwin: 'Sara thought it would be possible to send it to Lord Derby through Lady Derby. His would be a capital name if it could be got; but I don't want to give you any trouble; & unless you thought Mother could send it to Lady Derby nothing had better be done' (letter from W. E. Darwin to C. R. Darwin, [9 November 1879], Cornford Family Papers.)

Lady Derby's final letter thanks Darwin for a copy of his book on worms.

Knowsley, | Prescot.
16 Oct /81

Dear Mr Darwin

I am much obliged & greatly flattered by your kind thought of me. I have read your book with the greatest interest. You said once, laughing,—that you were finding that "Worms" could revolutionise the world;—you have succeeded in proving the greatness of their power.

I wonder how you fared at Down in the gale of Friday! We felt here as if we might be swept away. Seventy trees came down in an hour, people could not keep their feet. The storm was preceded by some minutes perhaps ¼ hour of perfect stillness—unusual stillness at 5. a.m. on Friday; the watchmen & others described "the roar as coming from the S.W for 3 or 4 minutes & then the wind burst in a hurricane".

I hear of great havoc at Holwood. Will you give my kindest regards to Mʳˢ Darwin & believe me | Yrs very sincerely | M C Derby.

The Nortons became friends of the Darwins after staying near Down in 1868. Susan Ridley Sedgwick Norton, wife of Charles Eliot Norton, and sister of William Erasmus Darwin's future wife, Sara, wrote this letter in 1871:

Dresden. | 9. Räcknitz Strasse—
Nov. 20ᵗʰ.

My dear Mr Darwin—

Truth compels me to state that I was not in search of pure science when I came across the little pamphlet which leaves here for Down tomorrow morning—far from it—but as I looked vainly, alas! for a french novel what should I see but the words "War Goethe ein Darwinianer"? Now I ask you, who are incapable of prejudice, if any better proof of German "fleissigkeit" [*diligence*] is wanting than that these admirable pursuers of hidden truth have actually time enough & to spare to steal the best genius of other countries?— Being in true feminine style convinced, without knowing anything about the matter that Goethe was no *Darwinianer*—I have not read the pamphlet—but Mr Norton has & he tells me that the profound Schmidt is of my way of thinking— You shall decide whether yr great original was to have been seen some time since wooing the lovely sirens of Weimar rather than those most interesting inhabitants of warmer climes—

Writing you this nonsense gives me a pleasant opportunity of telling you that we have heard from my Aunts & Sister, most animated accounts of your sons' visit to Cambridge [*Mass.*].— They have left behind them many friends & the most agreeable impressions & what more can one ask to do in going to a foreign country?— My Sister & brother imply that there was an immense amount of laughing done— So I take it that my country furnished at least one very admirable element of enjoyment—Mirth.— I wish we might have been at home to return a little bit of your unbounded hospitality to us but perhaps one of these days you may be fired with the desire to see those monkeys which one of yr great novelists describes as gaily gambolling in the trees of Illinois! If such should be the case you will surely not overlook Cambridge, the home of all virtue & learning & at least for a time will rest at Shady Hill,—where *novels* and a most affectionate welcome will always await you

You may be glad to know that we are most comfortably established in this dullest & most respectable of cities—& are all well—even Mrs. Norton [*her mother-in-law*] may be called well now—but

Germany is "langweilig" [*boring*] & I shall be glad when I find myself on the lovely shores of the dear little Island.

We send to you & yours warmest messages of regard—& Mr Norton bids me remember him very especially to yrself & Mrs. Darwin | Pray give her my love— | & always believe me | dear Mr Darwin— | Affectionately yours | Susan Norton.

> *Down,* | *Beckenham, Kent.*
> Nov, 23ᵈ

My dear Mʳˢ Norton

I am very much obliged for your kind & pleasant letter & for your present of the little book about Goethe. It is written by a very good zoologist, & I shall be glad, to look at it, but the German language is a sore grief & trouble to me.—

My sons enjoyed themselves wonderfully in America, & they met with really extraordinary kindness from many persons.— When I asked them what on the whole they liked best there, they answered without a moment's doubt "our stay at the Ashburners [*Susan Norton's aunts*] & the great valleys in California",—which seems an odd couplet. We heard much of all the fun & laughter they had with your sister & brother; & they heartily congratulated themselves that they had the good fortune to be invited to the house.—

I am glad to hear you are comfortably settled at Dresden, & as for dullness forgive me for saying that with your party it is not to be believed.—

Pray give my respect & kind remembrances to Mʳˢ. Norton, & good wishes to all your party, & I remain | My dear Mʳˢ Norton | Yours sincerely & obliged | Ch Darwin

P.S. My wife has just given me a good scolding, & I always tremble before her just severity, for not having given you her affectionate remembrances.—

In 1872, after the publication of *Expression of the emotions*, Darwin made contact with a very old friend from Shropshire, Sarah Owen, sister of Fanny Owen. Fanny had married and become Fanny Myddelton Biddulph. Sarah was now the widowed Mrs Haliburton, after a second marriage.

> *Down,* | *Beckenham, Kent.*
> November 1ˢᵗ

My dear Mʳˢ Haliburton

I daresay you will be surprised to hear from me. My object in writing now is to say that I have just published a book on the "Expression of the Emotions in Man & Animals"; & it has occurred to me that you might possibly like to read some parts of it; & I can hardly think that this would have been the case with any of the books which I have already published. So I send by this post my present book. Although I have had no communication with you or the other members of your family for so long a time, no scenes in my whole life pass so frequently or so vividly before my mind, as those which relate to happy old days spent at Woodhouse. I should very much like to hear a little news about yourself & the other members of your family, if you will take the trouble to write to me. Formerly I used to glean some news about you from my sisters.

I have had many years of bad health & have not been able to visit anywhere; & now I feel very old. As long as I pass a perfectly uniform life, I am able to do some daily work in Natural History, which is still my passion, as it was in old days, when you used to laugh at me for collecting beetles with such zeal at Woodhouse. Excepting from my continued ill-health, which has excluded me from society, my life has been a very happy one;—the greatest drawback being that several of my children have inherited from me feeble health.

I hope with all my heart that you retain, at least to a large extent, the famous "Owen constitution".—

With sincere feelings of gratitude & affection for all bearing the name of Owen, I venture to sign myself | Yours affectionately | Charles Darwin

My wife desires me to send her very kind regards to you.—

> Bridge House | Richmond | S.W.
> Nov.ʳ 3.ᵈ

My dear Charles Darwin

If I was to try & express to you, the extreme pleasure your letter had given me, to say nothing of the Book that accompanied it, I might be accused of flattery, or "soft Sawder"— Still, I may, (& I *will*) with truth declare that few letters, & few gifts have afforded me the gratification of yours yesterday— To know that I was still remembered by you, after such a lapse of years, is in itself a satisfaction

That remembrance has indeed been reciprocal, & often & often have I lamented that I never had a chance of seeing you— I have made enquiries from various friends of your's, & have always been told, that even the excitement of meeting an old friend, was usually more than you could bear— Spite of this, I will still indulge the hope of once more shaking hands with one of the best & most valued friends of my youth—

How many sad changes have befallen us both, since we met, & how many of those we most loved, have been taken from us,—to me, especially Life is but a shadow, a remembrance, of happy bygone days—

I have, like you, a most vivid remembrance of the bright old Woodhouse times, in which you stand first & foremost I can recall the Beetle, & the Fungus hunting, & above all, the glee with which "Charles Darwin" used to be descried, cantering up to the house, it being a received opinion, that any frowns of the poor governor would be at once dispelled, you being always the most influential favourite— I have now in my possession a letter you wrote to me from "Terra del Fuego", at my particular request, & I can often laugh at your boyish assertion, that the highest pitch of your ambition would be to be favourably alluded to, in Eddowe's Journal!—[*a Shropshire local newspaper*] I think that ambition has been attained, & *something more*—

I live a very quiet, solitary life, only associating with a few old, & kind friends, my house is pretty enough, actually on Richmond Bridge, with small garden sloping down to the River— My old passion for Animals still continues, but alas, I have no room for Poultry here, I have tried Pigeons, but they & the Cats were incompatible, so my live stock is now reduced to two tiny Maltese dogs, two very large persian Cats, & an old Cockatoo that I have had since 1848— I am certain your book will very much interest & amuse me— When one lives as I do, alone with Animals, their habits & manners become doubly interesting & familiar— My health is but indifferent, I fancy there is something amiss with my heart & the famed "Owen Constitution" is not what it was, in days of Yore. Time, & Sorrow, have much tried me—

You ask after the Family, poor Fanny, as you perhaps know, became a Widow 6 months ago, & now lives in London, with two unmarried daughters— Caroline Lister is settled in Yorkshire, Sobie [*Sobieski, another sister*] lives alone at Cirencester, Arthur now reigns at poor old Woodhouse, Francis lives on his small property near Overton, & Charles is Chief Constable of Oxfordshire, with a Wife & 5 Children— Of your Children, I have from time to time heard, that many of them inherit the Family talents, & I think you have one Daughter married, if not more—

When Summer returns, if we live till then, may I look forward to our meeting somehow & somewhere, I often go to London, & would meet you anywhere you might appoint though I must not run the risk of affecting your health—

Once more, thanking you from my heart, for the pleasure you have given me, believe me, always, most truly & affectionately Yours | S. H. Haliburton

Why did you address me so formally?—

Down, | *Beckenham, Kent.*
Nov. 6th.

My dear Sarah

I have been very much pleased by your letter, which I must call charming.— I hardly ventured to think that you would have retained a friendly recollection of me for so many years. Yet I ought to have felt assured that you would remain as warm-hearted & as true-hearted as you have ever been from my earliest recollection.— I know well how many grievous sorrows you have gone through; but I am very sorry to hear that your health is not good. In the Spring or summer, when the weather is better, if you can summon up courage to pay us a visit here, both my wife, as she desires me to say, & myself would be truly glad to see you, & I know that you would not care about being rather dull here. It would be a real pleasure to me to see you.— Thank you much for telling about your family,—much of which was new to me. How kind you all were to me as a boy, & you especially, & how much happiness I owe to you.

Believe me | Your affectionate & obliged Friend | Charles Darwin

Perhaps you would like to see a Photograph of me now that I am old.—

In 1880, Haliburton wrote again:

> *Bridge House | Richmond | S.W.*
> Novr 21st.

Dear Charles Darwin

(For I really cannot address you in any other way)

Yesterday I read, in a leading Article of the Times, "Of all our living Men of Science, none have laboured longer, or to more splendid purpose than Mr Darwin", & it recalled to my mind, your boyish assertion made many many years ago, that "if ever Eddowe's Newspaper alluded to you, as "our deserving Fellow Townsman", your ambition would be amply gratified"—

So you may believe with what sincere gratification, I see your fondest hopes, more than gratified, & realized— You have hosts of friends, but few older, or more sincere than myself, for you are associated with the happiest memories of my youth, & I have the most affectionate recollections of the name of Darwin, as connected with all that was good & pleasant— How my poor Father would have rejoiced in your "splendid success", & I can fancy his carrying that Newspaper about, & reading it to every body!—

It is a long time since I have heard any thing of you, but I hope you are tolerably well, as I see you are able to receive "Deputations"—

Let me hope we may live to meet again, meanwhile believe me always | Your's very affectionately | S. H. Haliburton

> *Down, | Beckenham, Kent. | (Railway Station | Orpington. S.E.R.)*
> Nov. 22^d 1880

My dear Sarah.

You see how audaciously I begin; but I have always loved & shall ever love this name.— Your letter has done more than please me, for its kindness has touched my heart. I often think of old days & of the delight of my visits to Woodhouse & of the deep debt of gratitude which I owe to your Father. It was very good of you to write. I had quite forgotten my old ambition about the Shrewsbury newspaper; but I remember the pride which I felt when I saw in a book about beetles the impressive words "captured by C. Darwin". Captured sounded so grand compared with caught. This seemed to me glory enough for any man! I do not know in the least what made the Times glorify me, for it has sometimes pitched into me ferociously.

I should very much like to see you again; but you would find a visit here very dull, for we feel very old & have no amusements & lead a solitary life. But we intend in a few weeks to spend a few days in London, & then if you have anything else to do in London you would perhaps come & lunch with us.

Believe me my dear Sarah | Yours gratefully & affectionately | Charles Darwin

My health is better than it was & I am able to do daily a good deal of work, but 24 hr^s never pass without some discomfort, & I am easily tired. Nevertheless there is much to make me happy & life is still an enjoyment.—

Bridge House | Richmond | S.W.
Decr 12th—

Dear Charles Darwin

It is no use! I cannot resist writing to tell you, what a real & great pleasure it was to me, to see you, & such a goodly Assemblage of Darwins besides, a gratification I had hardly hoped for— You are one of my oldest remaining friends, & you are so happily associated with the palmy days of yore, that it is indeed a heartfelt satisfaction to me to see you, & to feel assured, that old times are still fresh in your memory, & your friendly regard unabated—

I can only hope this satisfaction may be renewed at no very distant period for Life is short, & uncertain; But while it lasts, believe that I am always most affectionately your's, | S. H. Haliburton.

PS. | Our meeting had but one drawback, you called me "*M^{rs}. Haliburton*" twice— This offence must not be repeated—

Leith Hill Place | Dorking
Dec. 13. 1880

(Home tomorrow)
My dear Sarah

It was very good of you to write, & your note has given me much pleasure. It is not too common to find anyone in this world as true as steel. Your postscript is your own dear old self.—

Immediately that you left (Queen Anne St. Emma & I said to one another we must try when the weather gets a little better, whether she will face the dullness of Down & pay us a little visit. So that in the early spring you will have to make up your mind.

I had hoped to call & see whether Mrs Biddulph would admit me, & had got her address, but a Russian naturalist came to luncheon & dinned me half to death & then an American naturalist, & I was half dead. But next time that I am in London I will try. I think that there must be some Mrs Biddulph living in Leamington, for I was told so positively that our Mrs Biddulph lived there, that I have thought of enquiring. In former years I was, also, rarely fit to see anybody.

Let me call you | my dear old friend | Yours affectionately | Charles Darwin

Caroline [*Wedgwood, Darwin's sister*] is a little better & came down to dinner the first time for three months. She sends you her very kind love.

Haliburton's last known letter to Darwin, in 1881, contained condolences on the death of Darwin's brother, Erasmus.

> Pavilion Hotel | Folkestone
> Sep^tr 8th.

My dear Charles Darwin

I cannot refrain from offering to you my very sincere condolences on the loss of your Brother, for I know it must be a great sorrow to you, & I must always sympathize in all that concerns you— When I last saw Erasmus, this Spring, he appeared much in his usual health—

I am here, & have been, for more than three weeks, in attendance on my poor Sister Sobie, who returned from Aix les Bains in a most wretched state, & she has ever since been dangerously ill, Inflammation of the Lungs, & her Heart also affected, her recovery is more than doubtful, & her sufferings are sad to witness— Fanny Biddulph is here also, she desires many kind remembrances to you—

Believe me always | Most sincerely Your's | S H Haliburton

Do not take the trouble to answer this, I only wished to assure you, that your trouble had not been disregarded by me—

2   Marriage

Marriage was a pivotal moment in a Victorian woman's life. Women were often pitied because they only had the choice to say yes or no to a suitor (and sometimes, not even the choice to say no), rather than courting any man they liked, as a man might, in theory, court any woman. The character of a husband, how he meant to comport himself within the marriage, and the profession and family alliances that the wife would find herself involved in, were of vital interest. The status of a married woman was generally higher than that of an unmarried woman, and she might have more resources at her disposal; but she might also find herself at the mercy of a spendthrift or tyrannical husband, or unfriendly in-laws.

Darwin had four sisters (Marianne, Caroline, Susan, and Catherine) and numerous female friends and cousins, many of whom were diligent correspondents while he was on the *Beagle* voyage. His departure for South America coincided with an outbreak of weddings among his friends and relations, which was unsurprising given their ages, but must have left him wondering what, or who, would be left for him when he returned. The letters describing a series of weddings and their aftermaths are a wonderful source of information about mid-nineteenth-century courtship and marriage.

Shortly before he sailed, Darwin's friend Fanny Owen wrote in typically ebullient style about her sister Sarah's marriage to Edward Williams on 22 November 1831:

Woodhouse
Friday 2^d.—

My dear Charles—
    ... how I do wish you had been *with* us on the awful **22**^d. I am sure you would *thoroughly* have enjoyed it all—from beginning to end it certainly (tho' I say it who should not) did go off most brilliantly— I was the **Undertaker** and managed the whole affair from *cutting up* of a *Ton* of cake to making *gallons* of *Rum Punch* for the evening's festivities— Susan & I of course you know were the Bridesmaids, and Mr. Charles Jones the *Bridesmaid's Man*, about 10 carriages I think composed the Procession to Felton, the *dew Drops* **fell** about 11 o'clock, and I think really every body behaved with becoming fortitude & resignation—as for poor Mama she was wonderful   The *Bridegroom* I think was the most *flabbergasted* of the whole party, poor

thing I was quite sorry for him, he was as *white* as a *sheet* and as *I supported* him to the *halter* I really expected he w^d have fainted, no brandy was at hand tho' he entreated to have some— his stammering he was dreadfully nervous about but got through it all wonderfully the word *ch ch ch ch ch–erish* did stick some time but that was the only one— "As soon as the *ceremony* was over the happy Pair *stept* into their travelling carriage (*green* we are informed) and proceeded with all *possible speed* on a *romantic Tour* to the *Metropolis*, where it is understood they intend to attend all the Theatres"— you may laugh at this announcement but it is *a* **fact** — did you ever hear of any thing half so unsentimental, the very first night they arrived in Town off they went to Covent Garden, the next to the Adelphi, & so on every night they have been in London— it was so like Sarah—determined to lose no time— but to return to our festivities at home— we had actually 37 people to dinner, two Tables, I President of a side table, and *didn't* I pass the Champagne I never allowed a glass to be empty a moment but before the Cloth was removed all **my** *gentlemen* became so much *more elevated* than those at the other table that I began to be in a fright lest they should expose themselves—but luckily nobody was too much elevated except M^r. B. O. who rose to propose the *Ealth of our Ost* & *Ostess* and caused much fun— very numerous were the healths & toasts, drunk with 3 times 3—and several neat & appropriate speeches, also a beautiful song by that *wild Genius* M^r. Crofton, composed by himself for the occasion, and it is much admired— I w^d send it you but think it not worth a double letter— after dinner, the dining room was cleared & prepared for dancing. A most brilliant Ball we had, kept up with the greatest spirit till 5 oclock on Wednesday morning, all the servants dancing. Papa opened the Ball to the tune of "come haste to the Wedding", with M^{rs}. *Kenyon*, can you fancy them?— country dances were the order of the night, and excellent fun we had the only draw back was an occasional dreadful *kick*, from the *too well shod Fantastic* toes of some of the Beaux, but this was to be expected— altogether we had excellent fun, & I do **sincerely** regret my dear Charles, you were not of th⟨e⟩ Party, you might just as well have been as ⟨   ⟩ at Plymouth— How tiresome it is their keep⟨ing⟩ you so long in suspense—

On 20 December 1831, Darwin's sister Caroline wrote to describe Sarah's life as a newly wed. Sarah was clearly enjoying her status as a married woman:

> Susan & I sallied forth to pay our wedding call at Eaton, M^r. & Mrs Williams & M^r and M^{rs}. White having arrived there a few days previously— a few miles from Shrewsbury we met Sarah *alone in her carriage*. She was coming to see us, having for the first time she said "rung the bell" & ordered the carriage & she seemed quite proud of

having ventured upon such a step of authority— we got into the carriage to her & had a very merry drive with her to Shrewsbury— she made us but a short call & asked us to dine with her the next Monday which Susan & I promised to do which we did, & got there about an hour before dinner, we found Fanny & Sarah alone in the Drawing room. Fanny had we found insisted upon Sarah not running out to meet us in the Hall—but made her sit still "*to keep up her dignity*"— you cannot think how pretty & nice they looked when we came into the room, both looking so gay & happy— M.r & M.rs Bruce, M.r Edward, Hanmer, & Henry Hill, dined there— as soon as dinner was over Sarah & I sat together alone more than an hour & she was so open & affectionate that I hardly ever felt fonder of her. Edward Williams was in very good looks, thanks Sarah thought to a beautiful black satin neck cloth she makes him wear, but I think owing to his looking so very happy. I never before had talked much to him & was quite surprised to find how pleasant he was, & moreover he never once stammered— Sarah looked particularly lady like at the head of her table & did the honors very nicely. I believe there has been but one jealous fit since the marriage & Fanny said Sarah behaved exceedingly well— She was half frightened after telling us of it whether she had not done wrong, & I am glad to find from different things that they do not intend to talk of his faults now he is married to Sarah—

A little later Darwin's sister Catherine wrote to announce Fanny's own engagement. This was a touchier affair, since Fanny's fiancé had originally been interested in Sarah, and Charles himself had been thought to be interested in Fanny:

Shrewsbury.
January 8.th 1832.

My dearest Charles.

...

I must now begin and tell you Owen news, of which there is some very surprising, and extraordinary. Caroline spent two days at Woodhouse this week; she thought she should find them quite alone, & quiet, and what was her surprise, on entering the room, to find M.r Biddulph settled there.— You will be as much astonished as Caroline was, when Fanny took her out of the room, and told her that she was engaged to M.r Biddulph; he had proposed a few days before and been accepted, in the course of a secret ride, Fanny meeting him at the Queen's Head.— You may imagine how amazed we were, when Caroline came home, and told us; and I may add how grieved I was, when I thought of his dissipated, gambling character, though I am rather more reconciled now, as every body agree he is very affectionate, and *now* talks of spending great part of the year at Chirk Castle quietly.— I do not think Fanny cares for him half as much as she did for John Hill; but she is so exceedingly annoyed now, at the prospect of what Sarah will feel, when she hears it, that she thinks more about that, than she does about Mr Biddulph. M^r and Mrs Owen and all the family are very much alarmed about Sarah; they say she will be so dreadfully mortified, and so tremendously angry with Fanny, as she will of course fancy that Fanny was treacherous to her, and tried to attract Mr Biddulph, at the time of Sarah's flirtation with him. This is perfectly untrue, as Mr Biddulph declares, that his attachment to Fanny entirely arose from seeing Fanny's distress at the time of John Hill's desertion;—he was so charmed with her feeling and crying then, that he resolved he would try if he could not make her care as much for him; & from that time, his great anxiety was to shake off Sarah.— Mrs Owen is staying at Eaton now, to break it to Sarah; another and a great anxiety, as you may imagine, is that Mr Edward Williams should not find out that Sarah feels or cares about it. I think you will perhaps have a letter from Mr Owen by this Packet, as he said he would write to you, and I have no doubt, Fanny will write in it.— Your Portmanteau arrived safe the other day.— I must tell you that when Mr Biddulph first began to pay attention to Fanny, Mr Owen was so afraid that she should be "*blown upon*" again as Sarah was, that he woke Mrs Owen in the night, and declared, that if Mr Biddulph was only flirting again, he would call him out [*i.e. challenge him to a duel*] "and if I fall I shall leave my dying request to Owen, immediately to call him out again.—" Mrs Owen was much amused as you may imagine, at his midnight bloody thoughts.

January 29th. I must go on with my letter, and tell you all that has happened since I wrote; I know how much you care for Owen news. Mr Owen was the person who broke it to Sarah, and between being so sorry for her feelings, & so frightened, he actually cried very much, when he told her; however Sarah bore it very much better than any body hoped, and though she did not speak to Fanny for a day afterwards, she then shook hands with her. Mr Williams is happily kept quite in the dark, and Sarah really seems exceedingly attached to him, and he violently in love with her.— I don't think Sarah feels much about this marriage, beyond anger and resentment at Mr Biddulph, having made her a "*cat's paw*" last Winter; and of course she will get over that in a short time.— … Fanny's marriage is to take place in March, I believe. They go first to Chirk Castle for a week, and then to London for the Spring.— I have been staying at Woodhouse some days, while Mr Biddulph was there, and certainly thought him very agreeable, and I cannot help hoping that with such an attaching wife as Fanny, he will reform, and become tolerably domestic, and I am in pretty good hopes about that dear Fanny's happiness. You will find her a *motherly old married woman* when you come back.

I hope it won't be a great grief to you, dearest Charley, though I am afraid you little thought how true your prophecy of "marrying and giving in marriage" would prove.— You may be perfectly sure that Fanny will always continue as friendly and affectionate to you as ever, and as rejoiced to see you again, though I fear that will be but poor comfort to you, my dear Charles.—

…

Charlotte's letter will very much surprise you, as it has every body else. Only think of Charlotte's being going to be married after only a fortnight's acquaintance, with a man, who was a perfect stranger to all her family.

Charlotte's letter too will give you an account of Hensleigh & Fanny Mackintosh's marriage on the 10th of this month.— England is gone mad, with marrying, you will think. It is Leap Year, you know when the L⟨a⟩dies take *their* turn of proposing.

Charlotte Wedgwood (Darwin's cousin, and the sister of Emma Wedgwood, Darwin's own future wife) duly wrote describing the wedding of her brother Hensleigh Wedgwood and Fanny Mackintosh:

Dulwich

Jany 12— 1832

My dear Charles

… I cannot afford to let slip the opportunity of a wedding in the family as at the rate they have hitherto gone on at there is no chance of another occurring before you have finished your voyage round the world—as they are such very rare occurences it is a good plan that we are upon of doing up two members of the family at once tho on the other hand cousins marrying is a very humdrum affair & affords very little interest or entertainment— I am very much of Fanny M's maid's opinion, who being asked what she thought of her lady's marriage said "Well ma'am I think it wont make much difference". I am very glad it is over—they must feel so comfortable & at leisure now that the disagreeable interval since the time it was fixed is passed which was filled with nothing but tiresome settlings, moving of houses, & all sorts of plagues & Hensleigh between them & his new Magisterial duties was beginning to look worn out— he had been very unwell all the week & was in bed Sunday & Monday morning being to be married on Tuesday & he thought himself so bad that he had written a note, to say that the marriage must be put off a couple of days when luckily his doctor arrived, told him he was quite well, recommended him some mutton chops & wine which so restored him that his note was burnt, & he appeared at dinner time in very good condition for his execution next day—the only serious consequence of his taking to his bed was his writing to Fanny that she must get the wedding ring, an indignity that I should suppose had never been put upon a bride before— however she was obliged to submit & sent out one of the Thorntons for it— this was not the last indignity she was obliged to submit to neither, for first her gown did not arrive in time & she was obliged to strip one of her bridesmaids & be married in a borrowed one, & still worse she had to wait what seemed a long time in the church before the bridegroom made his appearance, & we began to be afraid he had taken to his bed again— however he appeared at last in very good case & accounted for the delay by his having a pair of hearse horses, 'a bad omen, & having to set down Judge Alderson at his chambers, & bad as this beginning was, for the rest of the time he cut a very good figure & he & the bride both took off their spectacles for the ceremony. There were eight carriages the servants with enormous favours which brought us a rebuke from the superior taste of an old dirty woman in the crowd, who said, "Well if she had been going to be married she would have kept those things out of sight & not collected a crowd about her" There was a grand breakfast afterwards at which Lady Gifford pre- sided 42 at table consisting besides all the branches of the family, of Thorntons innumerable besides a few other friends   Before we went to church Sir James [*Mackintosh: the bride's father*] made me stuff a vol: of a new novel into his pocket   I did not see what opportunity he had of reading it.

On 29 January she continued her letter with some possibly by now unsur- prising news:

Jan 29th My letter has been lying by a little more than a fort- night— how little I thought when I put it by what would be the next piece of intelligence that I should add to it—nothing less than that I am engaged to be married— I am afraid you will think part of what I wrote at the beginning of this very deceitful, but I do assure you it was not— I had not the least notion then of what was going to happen to me & that I should ever be married seemed to me the most improbable thing possible. You will have most likely heard this news in some of your other letters but you will like to hear more about it from me. When Emma and I arrived at Roehampton we found Mr Charles Langton staying there— he is nephew of the Mr Langton who married Marianne Drewe and is guardian to his son Bennet who always spends his holidays at Lady Giffords and it was to be with him during his holidays that Mr Langton spent this Christmas at Roehampton. He is a clergyman but has no living & has only a very small income now, but he was tutor to Lord Craven who has many livings in his gift & he has no doubt that he shall have one of them— he has also a rich grandmother so that he will be well off in future tho he is poor now. Some of Lord Craven's livings are in Shropshire & in very pretty parts of Shropshire—this will be delightful for me if Mr Langton ever gets one of them—to fall by chance so near home & Shrewsbury would be high good luck. For the present we are going to take a house in Surrey near Guildford—

it will be very pleasant to be within reach of London. Emma likes M^r L almost as much as I do & was delighted when he proposed to me, which I tell you because you will think her a more impartial judge than me. I looked forward to seeing you established in your parsonage but now I suppose I shall receive you first in mine. I think it is the happiest life in the world & I hope dear Charles that we shall hereafter compare notes upon it when we have both tried it & found it as happy or nearly as happy as we expect. In looking forward to it myself & thinking of its advantages I feel more anxious that you should finish all your wanderings by settling down as a clergyman but it must be as a really good active religious clergyman, (you know you gave me leave to preach) in that only can the happiness consist, & if I did not think Mr Langton would be all that, I think I would rather he were any thing but a clergyman. I feel a delightful trust in his high principles & kind nature which gives me a feeling of security that I have done what was wise as well as what was agreeable— it seems a very short time since I first saw him for me to judge so confidently of him & yet I do not feel the less secure for that. I am sure that in one respect being going to be married is very like going a voyage round the world—it makes one love all one's friends more than ever & it also makes one find out more affection in all one's friends than one ever knew of before—

Darwin, receiving the news in April at Rio de Janeiro, found this all rather overwhelming, replying to Caroline:

With yours I received a letter from Charlotte, talking of parsonages in pretty countries & other celestial views.— I cannot fail to admire such a short sailor-like "splicing" match.— The style seems prevalent, Fanny seems to have done the business in a ride.— Well it may be all very delightful to those concerned, but as I like unmarried woman better than those in the blessed state, I vote it a bore: by the fates, at this pace I have no chance for the parsonage: I direct of course to you as Miss Darwin.— I own I am curious to know to whom I am writing.— Susan I suppose bears the honors of being M^rs J Price.—[*John Price was a Shropshire friend of the Darwins.*] I want to write to Charlotte—& how & where to direct; I dont know: it positively is an inconvenient fashion this marrying: Maer wont be half the place it was, & as for Woodhouse, if Fanny was not perhaps at this time Mrs Biddulp, I would say poor dear Fanny till I fell to sleep.— I feel much inclined to philosophize but I am at a loss what to think or say; whilst really melting with tenderness I cry my dearest Fanny why I demand, should I distinctly see the sunny flower garden at Maer; on the other hand, but I find that my thought & feelings & sentences are in such a maze, that between crying & laughing I wish you all good night.—

Accounts of weddings soon gave way to reflections on married life. Catherine Darwin predicted, in a letter of 25 July 1832, that Fanny Biddulph would not have an easy life: 'the old Mother, Mrs Biddulph is so odious to her, and Mr Biddulph is such an exacting Husband.' In the same letter she suggested Darwin should marry Emma Wedgwood's sister Fanny:

> I hope you will in all probability find Fanny Wedgwood *disengaged* and **sobered** into an excellent Clergman's Wife by the time you return, a nice little invaluable Wife she would be; I will not quite promise though that you will find her disengaged, as another Clergman, Mr Paget Moseley, Brother to Mrs Frank Wedgwood, is said to be paying her very sedulous attention; but he is such a vulgar, fat, horrid man, I do not think it is possible she will have him.

(Sadly, Fanny Wedgwood died that same year.) On 27 September 1833, Catherine wrote of the Langtons' marriage:

> Charlotte seems extremely happy, very full of scrattles & household cares, (so unlike her.) Mr Langton is rather a talking, visiting, chattering man; whom no one would ever expect Charlotte to have fallen in love with; but so it is. The worst thing about him is to my mind, that he governs most absolutely in all little trifling concerns, as well as in great matters. Susan attributes this to his having been one year on board Ship when a boy, seeing absolute authority; if this is the case, what will become of your poor wife, after so many years apprenticeship in the art of governing?

Some interest began to be shown in the plans of Erasmus Alvey Darwin, Charles's only brother. Charles had predicted that Erasmus would marry Emma Wedgwood, and be 'heartily sick of her', as Catherine reminded him in a letter (*Correspondence* vol. 1, letter from Catherine Darwin, 26–7 April [1832]). On his return, he began to speculate that Erasmus would marry Harriet Martineau, who was at the beginning of her career as a writer, instead. This is from a letter written on 9 November 1862:

> Erasmus is just returned from driving out Miss Martineau.—[*i.e.*, *taking her for a drive*] Our only protection from so admirable a sister-in-law is in her working him too hard. … She already takes him to task about his idleness— She is going some day to explain to him her notions about marriage— Perfect equality of rights is part of her doctrine. I much doubt whether it will be equality in practice.

We can't tell how serious Charles's worries were, but Erasmus never married, and enjoyed literary and radical London society for the rest of his life.

Darwin himself became engaged to Emma Wedgwood on 11 November 1838, 'the day of days!' in his 'Journal' (DAR 158). The following letter of congratulation was written to him by Harriet Henslow, the wife of his mentor in Cambridge, John Stevens Henslow: Darwin had spent much time in their household as an undergraduate.

Cambridge
Nov. 22—

My dear Mr Darwin,

I do believe there are few pieces of intelligence that could have delighted me more than that which your letter of day contains, and you may be assured of the warmest congratulations of both Mr Henslow and myself— That your fit of *insanity* may long continue is my sincere wish, but pray don't let me hear a word of the *calamities* of domestic life, all such matters ought now to appear to you en couleur de rose, and if it is your Intention in future to resign *all such* cares to your fair Lady, it strikes me you do not intend to trouble *yourself* about them henceforward, *under such circumstances* I do not doubt you will prove a most *dutiful* husband, joking apart, I cannot help observing, what you are not to take as a compliment, because they tell me I never paid but two in my life, that I think Miss Wedgewood a very fortunate being; and that she has every reason to look forward to as great a share of matrimonial happiness as falls to the lot of most people. I am glad to hear it is your intention to reside in town, I feel that we have not quite lost sight of you, had you told me you were going to take up your abode in Shropshire, why then—I wonder if I should have read your letter with the disinterested satisfaction I ought to have done— Now I am on the subject of *Selfishness*, I must just say, that I do hope this event will not put a stop to the visit Mr Henslow has held out hopes of, of your coming to us at Christmas. I am sure you must require a great deal of information on household matters, so pray come, and talk it all over—

Believe me | Very Sincerely Yrs | H Henslow

Except for Elizabeth, all Charles and Emma's seven surviving children married: William, Henrietta, Francis, and Horace during Charles's lifetime. William and Henrietta in particular shared their thoughts about marriage in letters to each other and to their parents. William, the eldest, a banker in Southampton, wrote to Emma in 1869, the year in which he turned thirty:

Basset
March 4.

My dear Mother,

Thank you for your note, I am glad you had a successful stay in London. I have been trying to clear up my ideas on marriage for some little time past & find it a difficult job. I in within the last year or so the pros for marriage have appeared so disproportionate to the cons that I have hardly considered the matter at all; now, however, I think the cons appear to me to about balance the pros, therefore I am in a position to consider it on general principles without prejudice. So that in the first place I must take it for granted that what is natural & tolerably universal should have the benefit of the doubt & be taken as the right & wise course to go in for. But when one considers putting it into practise, the difficulties seem overwhelming. How is a man to get to know what a girl really is from meeting her at balls & parties? one sees nothing but society manners, society giggles, & society ideas all, entirely artificial, and made up to look well. Therefore Question No I is, is a man to go in like at Blindman's buff, & take the first that comes uppermost with about as much to guide him? Again granting one does in time get to know a girl to some extent, they all or a great majority of them turn out insipid with hardly any interests but in the most ordinary things, & none of them with any music, which would make up for a great pile of sins. Question no II is, is a man to go in just for an amiable sociable person and not much else, or to wait till he comes across something approaching his ideal? judging from my experience it comes to this, is a man wise in making the best choice he has the chance of, or is it better to remain unmarried sooner than marry any one but an exceptional person? Of course for simplicity's sake I omit just the practical difficulties, as they are secondary difficulties and only tend to intensify the former difficulties.

I should like to know what your ideas are on these points. It appears to me in one sense a misfortune to have a set of brothers & sisters & cousins decidedly intellectually above the average; it makes one too severe in one's appreciation of others' qualities, & probably too fastidious in one's ideal. …

Your affect son | W. E. Darwin

Emma replied on 6 March:

F.[*father*] wrote down the pros & cons long ago [*see pp. 30–31*] & if I can find the paper I will send it you. In the first place do not marry for marrying sake (look at Uncle Frank as a warning). What you say about the difficulty of knowing society girls is v. true. One knows a little what girls are from knowing what their parents are, that is to say if you liked what you saw, you w^d venture more safely than if you did not know the family. F. says marry a healthy wife & I say marry a wise one, which you will say is all v. fine talking.

Don't feel in any hurry, you are quite young enough. My opinion you know well enough that men are m. better & happier when young for being married, & that an old bachelor's latter life is generally desolate. Even Uncle Ras who is almost as fond of Hope & Effie

[*Wedgwood, Emma's nieces*] as if they were his own feels desolate &
melancholy.

When you do see anybody that rather hits your taste like that nice
Scotch girl follow up the acquaintance [*rest of letter lost*]

The first of Darwin's children to marry, however, was Henrietta. She married Richard Buckley Litchfield, a barrister almost ten years her senior
whom she met through friends in London. It was a whirlwind romance;
they met in June 1871, were engaged in July, and married in August. Emma
wrote in some bewilderment to her aunt Fanny Allen that it felt odd to her
that Henrietta should be so intimate with a person of whom she knew so
little. 'I feel quite at ease with him & that he is very nice, but I really have
not seen much of him.' (Wedgwood/Mosley archive, Wedgwood Archive
Collection, Keele University, cited in Browne 2002.) The wedding was
quiet, owing to Darwin's poor health, with no friends or relations invited
(a party came nevertheless from the Working Men's College in London,
where Richard taught). Darwin wrote to congratulate Henrietta while she
was on her honeymoon.

Down
Sept. 4th

My dearest Etty,

I must write to say how much your nice & affectionate letter from
Dover has pleased me. From your earliest years you have given me
so much pleasure & happiness that you well deserve all the happiness
that is possible in return; & I do believe that you are in right way for
obtaining it.— I was a favourite of yours before the time when you
can remember. How well I can call to mind how proud I was when at
Shrewsbury after an absence of a week or fortnight, you would come
& sit on my knee, & there you sat for a long time, looking as solemn
as a little judge.— Well it is an awful & astounding fact that you are
married; & I shall miss you sadly. But there is no help for that, & I
have had my day & a happy life, notwithstanding my stomach; & this
I owe almost entirely to our dear old mother, who, as you know well,
is as good as twice refined gold. Keep her as an example before your
eyes, & then Litchfield will in future years worship & not only love
you, as I worship our dear old mother.

Farewell my dear Etty.— I shall not look at you as a really married
woman, until you are in your own house. It is the furniture which
does the job. | Farewell | Your affectionate Father | Charles Darwin

Emma didn't agree about the furniture: in a letter to Henrietta in September 1871, she wrote, 'nothing marries one so completely as sickness' (DAR
219.9: 95). The Litchfields were indeed ill: Henrietta wrote from Cannes,
'R is a jewel of a nurse. We feel very married each lying sick in our beds as
if we'd been at it 30 years like Father and you' (DAR 245: 45).

Darwin's notes on marriage

First note [after 7 April 1838] (DAR 210.8: 1)

Work finished

If *not* marry | Travel. Europe, yes? | America????
If I travel it must be exclusively geological United States, Mexico
Depend upon health & vigour & how far I become Zoological
If I dont travel.— Work at transmission of Species— Microscope
simplest forms of life— Geology. ?.oldest formations?? Some ex-
perimets— physiological observation on lower animals
B Live in London for where else possible in small house, near
Regents Park—keep horse—take Summer tours Collect specimens
some line of Zoolog: Speculations of Geograph. range, & Geological
general works.—Systematiz.— Study affinities.

Work finished

If marry—means limited, Feel duty to work for money. London
life, nothing but Society, no country, no tours, no large Zoolog. Collect. no books. Cambridge Professorship, either Geolog. or Zoolog.—
comply with all above requisites— I could not systematiz zoological-
ly so well.— But better than hybernating in country, & where? Better
even than near London country house.— I could not indolently take
country house & do nothing— Could I live in London like a prison-
er? If I were moderately rich, I would live in London, with pretty big
house & do as (B), but could I act thus with children & poor? No—
Then where live in country near London; better, but great obstacles
to science & poverty. Then Cambridge, better, but fish out of water,
not being Professor & poverty. Then Cambridge Professorship,—&
make best of it, do duty as such & work at spare times—
My destiny will be Camb. Prof. or poor man; outskirts of London,
some small Square &c:— & work as well as I can
I have so much more pleasure in direct observation, that I could
not go on as Lyell does, correcting & adding up new information to
old train & I do not see what line can be followed by man tied down
to London.—
In country, experiment & observations on lower animals,—more
space—

Second note [July 1838] (DAR 210.8: 2)

This is the Question
Marry

Children—(if it Please God)— Constant companion, (& friend in old age) who will feel interested in one,— object to be beloved & played with.— —better than a dog anyhow.— Home, & someone to take care of house— Charms of music & female chit-chat.— These things good for one's health.— *but terrible loss of time.* —

My God, it is intolerable to think of spending ones whole life, like a neuter bee, working, working, & nothing after all.— No, no won't do.— Imagine living all one's day solitarily in smoky dirty London House.— Only picture to yourself a nice soft wife on a sofa with good fire, & books & music perhaps— Compare this vision with the dingy reality of Grt. Marlbro' St. Marry—Mary—Marry Q.E.D.,

Not Marry

Freedom to go where one liked— choice of Society & *little of it.* — Conversation of clever men at clubs— Not forced to visit relatives, & to bend in every trifle.— to have the expense & anxiety of children— perhaps quarelling— **Loss of time**. — cannot read in the Evenings— fatness & idleness— Anxiety & responsibility— less money for books &c— if many children forced to gain one's bread.— (But then it is very bad for ones health to work too much)

Perhaps my wife wont like London; then the sentence is banishment & degradation into indolent, idle fool—

It being proved necessary to Marry | When? Soon or Late

The Governor says soon for otherwise bad if one has children— one's character is more flexible—one's feelings more lively & if one does not marry soon, one misses so much good pure happiness.—

But then if I married tomorrow: there would be an infinity of trouble & expense in getting & furnishing a house,—fighting about no Society—morning calls—awkwardness—loss of time every day. (without one's wife was an angel, & made one keep industrious). Then how should I manage all my business if I were obliged to go every day walking with my wife.— Eheu!! I never should know French,— or see the Continent—or go to America, or go up in a Balloon, or take solitary trip in Wales—poor slave.—you will be worse than a negro— And then horrid poverty, (without one's wife was better than an angel & had money)— … One cannot live this solitary life, with groggy old age, friendless & cold, & childless staring one in ones face, already beginning to wrinkle.— …

Next came Francis Darwin, who married Amy Ruck in 1874. Emma sent an account of the wedding to his brother Leonard, who was travelling to New Zealand as part of the transit of Venus expedition. It seems to have been a calmer and better organised affair than the Shropshire weddings of the 1830s, but not so austere as Henrietta's.

Abinger
Wed | July 29.

My dear Leo

I hope nobody will have told you about the wedding. Bessy & I started at 8 a.m. on a lovely mg. Poor Old G. [*George Darwin*] did not feel up to going, which was a disappt to him. We got first to Q.A. S^t & found Frank in his work a day clothes & rushing about full of care, but what he was doing we did not know. W^m was there all prepared in his wedding trowsers which have been at so many. Ten minutes to the time we walked across to Langham church & just caught a glimpse of Amy running up the steps in her cloak over her gown.

They were lodging at Cav. St close by, so she & Mrs R. walked over. She looked m. shaken & agitated before the wedding; but after it was over as calm & comf. as poss. eating her own wedding cake, when we assembled at the lodging. She was in a white muslin & bonnet w. some of our Stephanotis in it. Dicky was prevented coming by some duty, & Edwal did not appear. I suppose nobody thought of wishing Ithel to come. Arthur was there. [*Dicky, Edwal, Arthur, and Ithil were Amy's brothers.*] We had a cheerful assembly at the lodging chiefly occupied in looking at Amy's beautiful jewelry, which w^d rival Lady L's if she put it on all at once in Lady L's fashion. [*Emma disapproved of Ellen Lubbock's style of dress.*] (The clergy. did the service nice & short, & seemed to me to excuse all the responses) Bessy & I came home with the jewels & George had great satisfaction in looking at them before putting them away in the strong box.

William himself became engaged in 1877, to Sara Sedgwick, an American. Sara was the sister of Susan Norton, wife of Charles Eliot Norton, a scholar, critic, and Darwin enthusiast. Susan and Charles had stayed in a village not far from Down in 1868; the two families had become acquainted and were very fond of each other. Susan died at the age of 34 in 1872. Sara's parents were already dead and she lived with her unmarried aunts, Anne and Grace Ashburner. On 24 July 1877, still nerving himself for the great step, William wrote to Emma:

Basset, | *Southampton.*
Monday

Private
Dearest Mother,

I went to Penn on Sunday and laid my mind bare to Hen. [*Henrietta*] She was very kind & sympathetic. I believe my mind is made up, and after a week's change of sense & ideas that I shall come back to Down determined to see her [*Sara*] again. I feel she would make me happy & I think I could her, though I should dread the dullness here of the society for her. Her health is poor, but I don't feel it much use in discussing that. It is odd I have often in former times had, when her name was mentioned, gleams of what a charming wife she would make.

Please keep secret as it is probably all moonshine & waste of resolution.

Goodbye dearest Mother, I am sure if such an extraordinary thing was to happen it would make you & Hen very happy.

…

Your affect son | W.E.D.

N.B. Don't understand this to mean more than that I am inclined to see her again

Despite his doubts, he proposed and was accepted. His sister Elizabeth thoroughly approved, writing the news to Ida Farrer, who was later to marry Horace Darwin.

> *Down,* | *Beckenham, Kent.* | *Railway Station* | *Orpington. S.E.R.*
> October 12th.

My dear Ida.

We have all been keeping a secret for the last 10 days but as it is no longer to be a secret on Saturday I shall take the liberty of telling to day. William is engaged to Sara Sedgwick! We are all very much delighted & have been able to think of nothing else. She is very nice indeed taking & pleasant, & friendly

She & William suit each other so very well, I cannot help feeling it a great pity that they did not see something of each other when she was in England 9 years ago, & perhaps they would have got engaged then. She is just about the same age as William nearly 38, she hasnt very good health but I think she is better now than she was.

She feels giving up her country & Aunts most dreadfully & doesnt mean to go to America before their marriage because she would feel it such a wrench to leave again.

Dear old William it is very nice to think of him being so happy, it is an odd thing whenever any body said who could William marry we used to say Sara Sedgwick but as she was miles away in America it seemed a little hopeless.

They are both coming here on Saturday, & she will stay some little time. As it was to be a secret William did the rather foolish thing of telling Uncle Ras, however he appears not to have told everybody as I should thought he would, but couldnt resist telling Hope [*Wedgwood*] that there was a blazing secret, but I daresay she guessed & that you all know, for it is extraordinary how things get round. William couldnt surprise his cook at all she said she knew it a week ago, & that she was very glad as you had to battle so to to make gentlemen understand things.

Emma wrote to Sara in January 1878, after the wedding:

Down, | Beckenham, Kent. | Railway Station | Orpington. S.E.R.
Thursday—

My dear Sara

It was a great pleasure to us both to receive your dear affectionate note & Mr Darwin sends you his love. It is a real happiness to think that you adopt us as something near & dear to you— Give my love to the Dragon. [*Sara's sister Theodora, who was evidently keeping visitors at bay.*]

I am glad to hear of her ferocity & hope she will be stern to the end. In London it so often happens that one returns home done up & then somebody calls to finish you— If you give orders before hand there can be no incivility in it, & then you have not the temptation & worry of thinking "Oh I sh^d just like to see such a one". ...

I think Theodora will be quite tame with us when she finds herself here unprotected by you, & we shall endeavour to treat her with every disrespect.

Yours my dear Sara | E. D—

The marriages in these letters cover almost fifty years of the nineteenth century. Despite the deepening seriousness of tone and the increase in decorum over the period, there are some constants. The most obvious is the anxiety of choice. Darwin and Wedgwood women regularly married men they knew very well, cousins, brothers-in-law, or life-long friends. When they chose a virtual stranger instead, the connections and status of the man concerned were used to reassure their nervous relations. Marriage within one's existing social circle was safe, but brought nothing new into the family. Marriage outside was risky and uncomfortable: a stranger brought into a tight-knit and exclusive family had to endure teasing, carping, and coolness, and so did his sponsor. But it could ultimately be rewarding. Both men and women worried about 'settling', and how much attraction was enough.

3   Children

Emma Wedgwood and Charles Darwin married in January 1839, set up a home in London, and celebrated the birth of their first child, William in December the same year. Over the next seventeen years, Emma bore nine more children (Annie, Mary, Henrietta, George, Elizabeth, Francis, Leonard, Horace, and Charles Waring), three of whom did not survive to adulthood (Annie, Mary, and Charles Waring). In 1842, the Darwins moved their family to Down, a small village in rural Kent, but within easy reach of London by railway. Emma and Charles ran a relaxed household, but were fairly conventional in their approach to education and health. Like many Victorians, they recorded details of their daily life, so anecdotes and observations found in diaries and notebooks combine with their letters to provide a picture of middle-class Victorian parents living in their semi-rural home. For Darwin himself, matters of family life merged seamlessly into research questions about the expressions of emotions and the early stages of human development.

Many middle-class Victorian families were preoccupied with health, and the Darwins were no exception. Charles's ill health is well known, but Emma also suffered regularly from illness, especially in association with her pregnancies and confinements. Details of Emma's pregnancy woes, from headaches to toothaches, found their way onto the pages of her diary, but it is clear from the letters she and Charles wrote each other that Emma didn't suffer in silence. Confinement was not an easy or safe event for Victorian women, and with the majority of births happening at home, it was very much a domestic affair. During the period in which Emma was regularly pregnant, maternal mortality in England and Wales was between 5.8 and 4.5 per 1000 live births, as compared with 0.082 in 2008 (Anderson 1990).

When Emma was pregnant with Mary and went to visit her family at Maer in Staffordshire, taking William (Doddy) and Annie with her, Darwin wrote to her from London on 9 May 1842.

Monday Morning

My dear Emma.

I am anxious for the post today to hear how you are & how the chicks are.— Yesterday felt quite a blank from not hearing— I hope your teeth have not been plaguing you & poor dear old Doddums temper I hope to hear is better.—

...

Emma Darwin with Leonard. DAR 225: 93.
By permission of the Syndics of Cambridge University Library.

After long watching the Postman your letter has at last arrived. you cannot tell how much I enjoy hearing about you all.— How strange poor old Doddy seems to be— I grieve he does not get better; I agree with you it w^d be very good to try calomel.— How astonishing your walking round Birth Hill, I believe now the country will do you good— What a nice account you give of Charlottes tranquil maternity— I wish the Baby was livlier,—for liveliness is an extreme charm in bab-chicks—

good bye.— I long to kiss Annie's botty-wotty | C.D.—

Darwin generally expected to be present for Emma's confinements: in 1856, he wrote to his friend John Maurice Herbert excusing himself from a trip to London: 'my wife is out of health & expects her confinement in a few weeks, & I cannot *possibly* receive anyone here or leave home' (*Correspondence* vol. 13, Supplement, letter to J. M. Herbert, 18 November 1856). Emma was pregnant with her last child, Charles Waring.

Charles administered chloroform to Emma during her deliveries possibly from as early as 1848, when Francis was born: this would have made the Darwins early adopters, since chloroform, pioneered by James Young Simpson in November 1847, was a controversial anaesthetic in obstetric and surgical cases. In 1850, Darwin wrote to John Stevens Henslow of Leonard's birth.

> Down Farnborough | Kent
> Jan 17th

My dear Henslow

… My said wife has been occupied these two days past in producing a fourth boy Darwin & seventh child! He is to be called Leonard,—a name I hold in affection from Cambridge & other associations.— I was so bold during my wifes confinement which are always rapid, as to administer Chloroform, before the D^r came & I kept her in a state of insensibility of 1 & ½ hours & she knew nothing from first pain till she heard that the child was born.— It is the grandest & most blessed of discoveries.

Yours affect | C. Darwin

Emma was reticent about writing the details of her confinements to her friends and relatives, and the fullest account of childbirth in the Darwin family letters concerns her daughter-in-law Amy. Amy was the child of Mary Anne and Lawrence Ruck and grew up in Wales with four brothers. The brothers met the Darwin boys at Clapham School. Amy married Francis Darwin in 1874. Francis had studied to be a doctor but gave up plans for a medical career to live in Down village and work as his father's secretary. In 1876, Amy gave birth to Bernard, Charles and Emma's first grandson. Emma wrote to Henrietta on 7 September.

Down, | Beckenham, Kent. | Railway Station | Orpington. S.E.R.
Thursday

Every thing quite prosperous—
Dearest H—

I had forgotten what a horrid thing confining was— Yesterday morning at 8. we heard that Dr Willey had been sent for. I went over about 10.30 to see what progress, & found Amy walking about & coming down to breakfast. She looked shaken & ill poor soul, but ready to talk about other things.

Dr W. came about 11.30 & said it w^d be a long affair but that every thing was quite right—so it went on all day— I backwards & forwards. Poor Dr W. made an attempt to go away & come again, but the nurse took a high hand & w^d not hear of it. I was going over for the 4th time about 8 when I met Frank coming about choroform, of which they gave her little whiffs to her immense comfort— She suffered most stoically & never made a sound— I came away at 10 & we went to bed having left orders to be informed when it happened. F. & I woke about 12.30 & began to get more & more uneasy at not hearing—when we heard some one at ½ past 3— It was William to say it was over & a boy—but what gave us a fright was to hear that he had fetched Mr Hughes—

I went over early this mg. & found Mrs Ruck writing telegrams as fresh as possible. She told me she thought every thing was q. right till 11. when Dr W. said he wanted to send for his instruments, that there was no progress making & they should have to resort to it at last when she was more exhausted. Mrs R. came over to our kitchen where W^m was sitting up & sent him off & was charmed w. his willingness & activity.

Mr Hughes came too, & I am so sorry they did not have him instead, Mrs R. says he was such a support & was a man of power, & Dr W. is only nice & gentle, (to be sure he is Welch & they talked Welch together.) He agreed w. D^r W's opinion & the child was soon brought into the world with a great cry.

The child is enormous so I am not surprized. Amy was not at all frightened, having read that is the fashion now to use instruments in America when every thing is quite right only by way of hastening the labour. When it was all over Mrs R. & Fr indulged in a good cry— And he went to bed utterly done up— He had been giving chloro. for 6 or 7 hours— Amy finding it an immense comfort but never insensible. Mrs R. comes out in the best light, so calm & cheerful & self forgetful.

I saw Amy for a minute owing to the nurse telling me she w^d be sure not to wake & I was a fool to believe her. She looked bright & happy & wanted to talk so I came straight away— She has been sleeping peacefully ever since— They say it is never likely to be necessary again. It was so fine I let the little Langtons & Smiths come all the same & Bessy undertook them & worked very hard & with

Amy Darwin. From Bernard Darwin,
The world that Fred made, 1955.
By permission of the Syndics of Cambridge University Library.

Lily & the nice governess they did very well playing out before tea & afterwards at the boat & slide upstairs which they enjoyed the most judging by the noise they made—

Yours my dear— … E. D

Friday

Dearest H.

Every thing quite right nursing & all— Mrs Ruck complains that is v. hard Amy *will* talk too much now for the first time in her life— She was not the least exhausted after all those hours. I went to see the baby—w. is uglier than usual, as I was sure a Ruck baby wd be— It seemed uncomf. & was sick v. often & I found to my surprize that the nurse had given it gruel— I felt it so dreadfully dangerous for one grandmamma to fight the other about her favourite nurse, that I only expressed a mild doubt whether gruel was suitable food for a new born child, & I was v. glad afterwards that I said no more, as I found Frank had flared up manfully & quite forbidden it again (I am surprized to see how. m. of a D^r he is—even to the loss of every vestige of decency)—

Bessy was not to see Amy, so I persuaded her to fulfil her engagement to Emily & if everything is smooth she is to bring her home tomorrow to spend Sunday— …

yours my dear— | We have had wet but pretty days—

Amy died on 11 September, after eighteen hours of convulsions. The two families were devastated. It was decided to bury her in Wales. Emma wrote to William on 13 September.

Wednesday

My dear William

You will like to know all I can tell you. Yesterday George & Horace were in London arranging for the trains for the conveyance of the coffin, so as to have no changes & they managed it with some difficulty by going from Bromley to Willesden— Horace slept in Q A St so as to be ready to travel down w. Mrs R. Frank & Arthur today— Dickie & Mrs Atkin [*Amy's sister*] go down with the coffin tonight. The last thing I saw of Frank was his coming in last night just to see the baby— I am afraid he will be more utterly miserable than he is now, as he has violent bursts of tears & overflows with affection for us all— He has asked his father to send proof sheets to him to copy & he has taken down F's autobiography to make a copy of it—which will be work without much thought. He does not mean to attend the funeral which is to take place at a beautiful little Church 5 miles off.

He took a pleasure in looking often at poor Amy & adorning her with flowers & came over in the evening after it was dark to get some

Stephanotis & it seemed to add to his bitter grief not being able to find any but one spray— They went off at 6 this morning calmly Parslow said—

Mrs Ruck is his great support— She is always able to speak. It all came a sudden blow on the poor brothers & I never saw any thing like their grief— Afterwards they were most thoughtful & useful—

Poor Bessy can only sleep very little & is utterly shattered. She feels truly that she can never hope to have the loss of Amy replaced   She was so sympathetic & the only person B. could be open with. F. is distracting his mind with schemes about building an additional room so that Frank may be made comfortable. I hope George will run down & see you very soon—

We shall wait till we have established the baby with a wet nurse— when I am sure Hen. w^d come & stay with it while we were with you— They came from Glasgow yesterday morning   they heard the bad news there & perhaps it was better than the suspense. She found the car rough & swinging very contrary to your experience but I think it has done her no harm.

My heart aches whenever I think of Frank; but now he is out of our sight we shall be able to forget him more & take to our usual occupations. F. has borne it wonderfully & I am quite well—

My dear William I long to see you— Do you keep as well as you can   E.D.—

Charles and Emma took on the care of Bernard and of Francis, who moved back into Down House with his parents. By now Charles and Emma were confident with babies and small children. Their early experiences as parents may have been a little more uncertain. Emma's relaxed style of housekeeping and childcare did not sit well with her Darwin in-laws. When Darwin took their first-born, William, to see his father and sisters, leaving Emma behind in London, he received a barrage of advice, which he (possibly unwisely) passed on, in a letter of 1 July 1841.

I am grieved to hear my Father, who is kindness itself to him, thinks he looks a very delicate child— He says the cough proceeds from the stomach; but he cannot feel any hardness in it & he has felt it well.— He thinks the iron & chalk of the greatest consequence to him, but decidedly injurious if his bowels are not fully opened— He says he has no doubt the rhubarb [*used as a laxative*] in the cakes has been injured by the baking— I am most glad we sent him here— I have picked up even already many hints— he particularly wishes him to have plenty of meat.— & I felt quite ashamed, at finding out, what I presume you did not know anymore than I did, that he has had *half a cup of cream* every morning—which my Father (who seemed rather annoyed) says he believes is one of the most injurious things we could have given him— When we are at home, we shall be able to look more after him.— Only conceive Susan found him when he started in the carriage with his stocking & shoes half wet through.— My Father says getting his feet wet on the grass if afterwards changed is rather a good than bad thing, but to allow him to start on a journey in that state was risking his health— Last night Susan went into Doddy's room & found no water by his bed-side— I tell you all these disagreeablenesses; that you may feel the sam⟨e⟩ necessity, that I do, of our own selves look⟨ing⟩ & not trusting anything about our children to others— I most heartily pray my Father may make poor dear Doddy look more robust & lose his cough.—

Emma did not reply.

Emma and Charles worried about the health of their children throughout their lives, and their children relied on their parents for comfort and advice. Henrietta recalled her mother as excellent caregiver:

> My mother's calm strength made her the most restful person to be with I ever knew. To the very last it was always my impulse to pour out every trouble to her, sure that I should have sympathy, comfort, and helpful counsel. She was a perfect nurse in illness. Her self-command never gave way and she was like a rock to lean on, always devoted and unwearied in devising expedients to give relief, and neat-handed and clever in carrying them out. (*Emma Darwin* (1904) 2: 5)

Charles also cared for his children when they were ill. Henrietta recalls sick children resting on the sofa in their father's study while he worked 'to be quiet and safe and soothed by his presence' (*Emma Darwin* (1904) 1: 468).

Middle-class Victorian families tended to be large: Charles and Emma's was only slightly above the average for families of their sort, which was about 5.7 to 6.2 children per married woman (Anderson 1990). They were also not unusual in suffering the loss of children for various causes. Mary died just over three weeks after she was born. Emma wrote to her sister-in-law Fanny Wedgwood a few days later, on 20 October 1842.

> Thank you, my dearest Fanny, for your sweet, feeling note. Our sorrow is nothing to what it would have been if she had lived longer and suffered more. Charles is well to-day and the funeral over, which he dreaded very much. … I think I regret her more from the likeness to Mamma, which I had often pleased myself with fancying might run through her mind as well as face. I keep very well and strong and am come down-stairs to-day. I have had a good acct from Maer. …
>
> With our two other dear little things you need not fear that our sorrow will last long, though it will be long indeed before we either of us forget that poor little face. Goodbye, my dear Fanny, with my best love to Hensleigh. | E.D.
>
> Every word you say is true and comforting.

In 1850, the Darwins' eldest daughter, Annie, fell ill. In March 1851, other measures having failed, Charles took her to Malvern with the children's nurse, Jessie Brodie, to see if the water cure would improve her health. A few days later they were joined by Henrietta and the governess, Catherine Thorley, and Charles returned to Down. By April, Annie had taken an abrupt turn for the worse. Emma was heavily pregnant and could not go to her: Thorley, Brodie, and later her sister-in-law Fanny Wedgwood had to act as her proxies. As Annie's health took a downward spiral, Catherine wrote in detail to Emma.

My dear M^{rs}. Darwin,

Dr. Gulley's opinion now is that Annie is very slowly progressing; this has much relieved my mind. He came early today (his cold was too bad to allow him to come last eve) & I again asked if he thought there was danger. "No," he said, "it is a smart bilious gastric fever, but she has turned the corner," these are the words he said he has many similar cases on hand just now, & he said it is quite an epidemic; hers has not been brought on by her treatment with him, but it is always more or less general at this season particularly when easterly winds prevail, of which we have had no small share for the past week.— On recovery I think debility shows itself in an invalid more than at the time of fever, for we now discover her extreme weakness; Dr. G. allows to-day a tablespoonful of broth of Beef Tea every hour; sickness has returned yesterday & the day before in the afternoon, so sh^d it do so this aft. we are to omit the broth for an hour or so. To-day she has scarcely complained of sickness so I do trust that is gradually disappearing & then we can venture to give something more substantial even than broth. Dr. G. considers that this will remove much of her Chronic Disease; this I was to inform M^r. Darwin. Brodie is in much better spirits to-day; I must say that I had serious thoughts yesterday morn. of sending for M^r. D., she was so very ill; I hastened down to M^{rs}. Scott to learn what she thought & there w^d have been an opportunity of sending as she is gone to town this day; for I almost feared my note of yesterday w^d. not reach you before this does but I was thankful to be able to give her a cheering account before her departure this morn. Annie has been obliged to have an Injection, as they never give purgatives in Homœopathy.— M^{rs}. Panting called again, she is truly kind & sympathizing.— Etty has gone out for a walk with Brodie; she is very well.— I trust M^r. Darwin is improving & that I have not alarmed you both in my communication of yesterday; as I trust now all the worst is over, I almost hope that Sunday's letter may not reach before this to-morrow.— Dear Annie sends her love; how I long for the time to be able to say she is strong enough to get up.—

Adieu, ever yours, my dear M^{rs}. D. | C. A. Thorley

Montreal House | Monday 4 oclock P.M

43

Dear Annie is a shade better, my dear Mrs Darwin; tho' Dr. G. says she is not out of danger; still he sees many good signs; in fact he says he has seen many children younger than she is where they have been in a state of stupor for three days, & yet they have recovered; Annie's excitable temperament is his only reason to make him fear. We passed a most anxious night but she got over 3 o'clock A.M. very well which Dr. G. told me was the critical time; Dr. G. thinks her decidedly better to-day, he has been twice; & comes again at 7. we are giving her now a dessert spoon of white wine every hour, & a medicine he prescribed last night, these two things she has taken for the last 18 hours; the wine wrought wonders at first, but of course that great effect must fail after a few doses.— I have just enjoyed feeding her with some orange juice which the dear child thoroughly relished. I long to see Mr D. tho' we fear he may be very much affected by the sad intelligence of yesterday. Where there is life there is hope & I cannot help hoping even more than this morn. particularly, as she has passed 3 o'clock P.M. so well; which was the time she became worse yesterday.— Dr. G. is truly kind, it is impossible to be more so. Poor Brodie feels it deeply but still bears up. I gave way sadly on learning first Dr. G's opinion but I am wonderful to-day. Dear Etty is amusing herself with her new doll, beads, &c. she is very anxious about the dear child Annie. I am sure you must be most desirous to be here, but rest assured that all I can any way do for the dear girl, has been & shall continue to be done.— There is nothing more I can say of Annie as she has been about the same all the morn. dozing a good deal & her mind wanders at times but this is not to be wondered at from the quantity of wine she has taken, all fever is removed which is a great thing. Make any use you please of Mama's offer as I know she will be glad that you do as it suits you best. there will be a room for Mr D. I trust you keep pretty well.

With Etty's love to all | Ever yours | C.A.T.

With Emma weeks away from another confinement, Charles rushed to Annie's bedside. Henrietta recalled her father's arrival on 17 April, 'flinging himself on the sofa in an agony of grief' (*Emma Darwin* (1904) 2: 141). Emma asked Fanny Wedgwood to join him. Such mutual care-giving was not out of the ordinary. When Emma's brother Hensleigh Wedgwood fell ill in 1842, Emma brought Hensleigh and Fanny's three children to Down (*Emma Darwin* (1904) 2: 51). As Annie slipped away, Emma corresponded furiously with Charles, Catherine, and Fanny. Multiple letters were sent each day, and Emma recorded Annie's symptoms and treatment in her diary. This letter is from Fanny Wedgwood.

Saturday | 7 oclock

My dearest Emma.

Charles tells you every thing of your darling child but you will like to hear any other impressions— we are waiting for D^r G's visit   she has been sick again since 4 oclock when Ch closed his letter but is looking more comfortable & has seemed quite to like being turned on the other side. her temperature is right & pulse about the same, but it is quite surprising that she should bear the vomiting so much better than she did 3 days ago I do not think her so emaciated as I expected, not so much I am pretty sure as Bro [*Fanny and Hensleigh's son*] was in his illness she has looked about more today & her face has strong eyes a more natural expression—extreme languor & prostration but no oppression about the head or eyes. she has just asked Miss Thorley quite loud something about her watch, but much of what she says we cannot make out from the roughness of her poor mouth—but I do generally make out the 'thank you' almost always— It is a blessing to see no suffering from the vomiting does not seem to be any now— dearest Emma how thankful I am to be able to be of least use to Charles. he looks really not ill though sometimes of course most sadly overcome & shaken   he has been two little walks today   I do not try to prevent him doing a good deal about dear Annie it seems as if it was some relief to be doing something, tho' occasionaly it may be too much There is nobody to be pitied till you, but I know how much you wish to be able to bear for Charles' sake

May God grant you both the life of y^r child

your Ever most affec^{ly} | FEW.

¼ to 8 | D^r. G. is now here there is a decided improvement in the tongue—a most important point. pulse quick & sharp still but perhaps from a little brandy. Dr G. thinks the vomitting not so unfavorable—it is better under the circumstances than the matter being absorbed

Down
Tuesday | before post time

My dearest Fanny

I must thank you for writing so fully & for telling me about Charles. Your impression of our poor child's looks was a comfort. Your being there is an immense comfort to Charles & I think you are quite right to let him do as much as he can as it must be the greatest relief he has. If he should be ill your being there will be a double blessing.

I feel today very awful being the end of the fortnight. I was sorry I sent for Etty when I heard of the pleasant scheme for her, but Mrs Rich & Snow [*Frances Julia Wedgwood*] had not the heart to disappoint her so she was off before Bessy arrived & I am quite satisfied.

Goodbye my dearest | God bless you for all your kindness | ⟨E⟩ D

Annie died on 23 April. Her loss had a lasting impact on Emma and Charles, and on Henrietta. Annie had been undeniably a favourite child, and Henrietta endured a painful sense of guilt and inferiority after her death. Three more children were born, Leonard, Horace, and Charles Waring. Henrietta herself became dangerously ill. Henrietta's manuscript recollections, and her biography of her mother, reveal an undertow of resentment at the burden of her mother's frequent pregnancies.

Despite these traumatic episodes, it was generally a happy, noisy and boisterous family. The Darwin children had free access to the house and grounds. Their cousin Snow Wedgwood once remarked that 'the only place you could be sure of not finding a child was in the nursery' (*Emma Darwin* (1904) 1: 468). George recalled a loud and lively household that didn't stop while his father worked:

> However hard my father was at work we certainly never restrained ourselves in our romps about the house, & I sh^d certainly have thought that the howls and screams must have been a great annoyance; but we were never stopped. There was one fearfully noisy game which invaded the whole house called "roundabouts" & we generally played at this when there was a house full of cousins. It was a modified hide & seek & necessitated yells from all the players to tell where the demon of the game was. As an example of the case of our fooling with my father I may mention that a game which Henrietta & I enjoyed much was to sit together in his microscope chair with a walking stick. (DAR 112: B11)

Francis recalled that the garden at Down was 'originally a bare and windy wilderness', which their parents rectified by building up mounds of raw red clay. The children referred to these mounds as mountains and would cross the passes on stilts. Emma often read and drew with the children. Many of her diaries contain scribbles and doodles by both her and the children, and the back of Charles's manuscript pages were often used for sketching. At times Emma joined in the children's play. Francis recalled that she 'laboriously reddened with sealing-wax' a regiment of dragoons to convert them into British soldiers (F. Darwin 1920). She was also known to take a slide down the sliding-board that they put on the stairs, writing to William in February 1857, 'Mr Lewis has made a sliding board for the children & they enjoy it very much. They put it on the stairs & I have taken a slide or two & so has Miss Thorley' (DAR 219.1: 16).

The children played a wide variety of games among themselves and with their parents. Sometimes the games were as simple as billiards, the table being a substantial purchase in 1858. Emma began a letter to William on 12 March 1859: 'George came yesterday & you may imagine his devotion to the billiard table. Papa thinks he will play well. "Train up a child" &c. All his spare time he devotes to knuckle bones so that he is not intellectual just now' (DAR 210.6: 39). Five years later devotion to billiards was still strong, with Emma ending a letter to William in May 1863: 'Horace is certainly better & devoted to billiards as he cannot have his beloved swing. He tires out Hen. [*Henrietta*] & Papa' (DAR 219.1: 74).

When it came to education, Emma and Charles were conventional, except that Charles chose a scientific rather than a classical education for his sons after William's unsatisfactory experience at Rugby. Emma and a series of governesses were largely responsible for the children's early education, and the girls' later education. Henrietta read voraciously and later became one of her father's editors, but it was Elizabeth who, in 1862, asked to be sent away to school. Her parents obliged. Emma wrote to William: 'We were thinking of sending Lizzy to school so as to get a month of it before the holidays. She is very stout hearted about it & is learning to mend stockings &c.' (DAR 219.1: 64). Elizabeth started at a boarding school in Kensington run by Miss Buob in January 1863. Emma shared Elizabeth's reactions to school with William:

> Lizzy's first letter began very cheerfully but ended rather low about her spelling & being so much more stupid over her lessons than she is at home & having a theme to write which does seem to me a very difficult think for any body to do, about as difficult as a sermon. I hope she had an outing at Cumberland P. with Rose & Mabel [*daughters of Emma's brother Frank*] on Sunday & we shall have her at Q. Anne S! next Sunday.

Her letters to her sons whilst they were at school showed a similar concern for their education and their wellbeing. In March 1852, Emma wrote to William at Rugby School.

Down
Tuesday

My dear Willy

Papa & I were delighted with your tremendous long letter & I don't wonder it made your fingers ache. It told us just what we wanted to know & began so regularly from the beginning. I shall not be able to write you such a long one, for we have gone on so quietly there is very little to tell you. I think Mr Mayor must be a good natured man to take so much pains to make you comfortable at first. Papa said school was a very different thing when he was at Shrewsbury school & nobody would have troubled themselves as to a boy's sitting at dinner or sleeping at night. Mrs Morrey has been very unwell last week & Mr Morgan has been to see her several times & bled her in the side. She is better now & gets up. Georgy is more crazy about drawing than ever & makes little picture books for Franky & Lizzy which I buy from him to give them, but he draws such a number that he does not take pains & so does not improve. D^r & Mrs Hooker do not come as he is too busy, but I hope the Chester Terrace folks & Aunt

Eliz. will come tomorrow all the same. Poor Wasp was sent back to his master & came back again by himself one night, but Parslow took him back again. It was lucky you had that large trunk as it will be more handy to keep your clothes in than a portmanteau would have been. I shall tell aunt Susan about the poor little sweep as she will be interested in it. I hope he will not be so foolish as to be enticed away again. Give my love to Erny [*Ernest Wedgwood, a cousin*] & with Papa's love to you   I am my dear Willy | your affectionate mother | E. D.

I never understood about the studies till I saw your plan.

We shall be very anxious to hear where you are placed. When you write to Mr Wharton [*William's tutor*] do not write a very short letter as that does not seem friendly, but tell him about your study & being with Erny &c. & send your love or kind remembrances to Mrs Wharton.

I [*don't*] know whether you have got your Mrs Markham's England [*A history of England from the first invasion by the Romans to the end of the reign of George III*] but if you want it we will send it by post.

Like Charles, Emma had concerns about education and she was not afraid to take public action when she saw fit. Corporal punishment was not uncommon when the Darwin children were in school, and if used infrequently was not seen as unjustified. Emma wrote to William on 24 November 1852:

> Down
> Wednesday
>
> My dear Willy
>
> I am glad you told us about your caning. I should like to know what you failed in, & what is being floored? it sounds like being knocked down, but I don't think Mr Burrowes looked fierce enough for that. I hope you will make up for it by working very hard now & I have some hopes you may be removed [*put into a higher form*] as your character has always been good.

However, in 1884, Emma took issue with the Down schoolmaster's approach to discipline and punishment, writing to George on 17 August:

> I am engaged in warfare with Mr Ducker the schoolmaster, having discovered that his habitual method of teaching is very brutal; caning pretty severely for a wrong sum bad dictation or blotting a copy book— He has been reprimanded several times & is under an engagement to put down in his Log book every punishment. This he has neglected to do; but £.s.d. comes into the question, as he keeps up the reputation of the School & gets a grant, & so he will not be dismissed this time.

I wrote to Sir John [*Lubbock*]; but he said in answer that he was so little at home he c^d not judge. Mr Forrest is very zealous, & is to call a meeting & insist on Sir John coming. The next time any flogging for such causes takes place I shall refuse to pay the school rates, if he is not dismissed, & Mr F. says he shall do the same— I am very friendly w. Mrs Ducker who teaches Bernard, wh. makes it rather disagreeable.

4   Scientific wives and allies

When Darwin made notes in 1838 about what he intended to do with the rest of his life, the question of whether or not to marry was a critical one. 'If *not* marry', he mused, 'Travel. Europe, yes? America????'. 'If marry', he continued, further down the page, 'means limited, Feel duty to work for money. London life, nothing but Society, no country, no tours, no large Zoolog. Collect. no books.' On another sheet, he wrote with qualified optimism of the advantages of marriage: 'Constant companion, (& friend in old age) who will feel interested in one,— object to be beloved & played with. — —better than a dog anyhow', while still noting the 'terrible loss of time'. On the side of the single life, he wrote, 'Freedom to go where one liked— choice of Society & *little of it.*— Conversation of clever men at clubs— Not forced to visit relatives, & to bend in every trifle.— to have the expense & anxiety of children— perhaps quarelling— **Loss of time**.' The risk that marriage would distract him from his scientific interests seems to have been uppermost in his mind, but at some point a sort of emotional spasm occurred: 'My God, it is intolerable to think of spending ones whole life, like a neuter bee, working, working, & nothing after all.— No, no won't do.— Imagine living all one's day solitarily in smoky dirty London House.— Only picture to yourself a nice soft wife on a sofa with good fire, & books & music perhaps'. (DAR 210.8: 1 and 2.)

Shortly after that, Darwin proposed to Emma Wedgwood and was accepted.

Darwin was right to see marriage and his future career as inextricably entangled, but seemed unaware that his family could be an asset rather than a liability, although that in fact was the case for him and was a fairly normal state of affairs among contemporaries of his own rank in life: unless of course he was joking. His notes on marriage seem bleakly comic. The Darwins and Wedgwoods were so steeped in Jane Austen that it's difficult to imagine Darwin writing them without an occasional wry smile at himself. In fact the wives of scientists, and sometimes other female supporters, as well as providing the domestic support that was expected of them, acted as secretaries, research assistants, translators, and fund-raisers: and, as mistress of a house, a scientific wife could have an important role in what we would now call networking.

Darwin's early letters to Emma dwell on the anticipated pleasures of domesticity and the comedy of his preparations for it (to be fair, he was the one with the primary responsibility for finding and furnishing a house in London: propriety made it impossible for Emma to spend much time with him before their marriage). This letter from Emma to Charles of 21 November 1838, before their marriage in early 1839, however, shows Emma's sense of her responsibilities as the wife of a man of learning. He had evidently asked her to dispose of some geological specimens.

> Maer
> Wednesday
>
> My dear Charles
>
> I am afraid the D^r [*Darwin's father, Dr Robert Darwin*] would not think it according to the strictest etiquette my writing to you before I have received your letter, but perhaps my having had a letter from you from Shrewsbury may save my dignity a little. When you were gone we all went a walking & for the sake of shelter happened to go exactly the same way that we went in the morning & as I don't set up for dignity I may confess to you that I felt it rather a contrast & was very stupid & flat all day. You might well say that there is a fate about specimens as you shall hear. I took your stones out of my cupboard in order to perform the first solemn duty you had ever imposed upon me but forgot them & left them wrapped up on the drawing room table & I was much alarmed on my return home to find Uncle John sitting there with them. I carried them off & threw them into the pool & just when they were past recal recollected "suppose he should have seen them & want to look at them again or have them back again." However it is all safe & nothing has been said. ...
>
> Goodbye my dear Charles yours most affecly | Emma W.
>
> You will kindly mention any faults of spelling or style that you perceive as in the wife of a literary man it w^d not do you credit, any how I can spell your name right I wish you c^d say the same for mine.

In a similarly satirical mood, Darwin wrote to Emma on 20 January 1839 of a visit to Charles and Mary Lyell:

> I was quite ashamed of myself to day; for we talked for half an hour, unsophisticated geology, with poor Mrs Lyell sitting by, a monument of patience.— I want *practice* in illtreating the female sex.— I did not observe Lyell had any compunction: I hope to harden my conscience in time: few husbands seem to find it difficult to effect this.—

At this time Darwin and Lyell had a running joke in their correspondence about their plan to abandon their wives once Darwin was married and spend their evenings at scientific clubs. Mary Lyell was a learned woman (see chapter 7), and if she was bored it can only have been because Darwin's geology was more elementary than hers.

In any case the image of men escaping the home in order to talk about science in clubs turned out to be inaccurate. The Lyells' house in London became a centre for scientific society. Mary Lyell's importance as a hostess is shown in this letter to Darwin from J. D. Hooker, 2 June 1865. Hooker had been discussing a bitter dispute about suspected plagiarism between Charles Lyell and John Lubbock.

> And now my dear D. shall I tell you what is at the bottom of it all?—perhaps you wont believe it— it is just this—that Lady Lyell will not call on Mrs Busk nor invite the Busks to her parties. This the Lubbocks' & Huxley's resent. You never agreed with me about the Lyells position respecting their Scientific reunions—but I always told you they were playing with the fire, & would assuredly burn their fingers.— Here is Busk, an FRS, a Secy of Linn. Soc.—Hunterian Profᵗ.—elected by Committee in Athenæum Examiner to the Army Medical Board, & God knows what all, besides being a universal favorite—is called on by Lyell to be pumped dry of his knowledge; living in the same street for years with the Lyells', & never otherwise noticed by them.— His wife, a most thoroughly accomplished clever person, excellent wife & mother, really scientific, & the kindest & most hospitable charitable person alive, more of a Lady than others asked to Lady Lˢ. soirées— I do say that the Busks' must feel this to be social ostracism & nothing else. It is all very well to say that an Englishman house is his castle, & that it is no one elses affair who is invited to the house & so forth—but Lady Lyells Soirees are quasi public.— Every Englishman & foreigner of distinction—friend or stranger is invited, & the Owens & Carpenters & Busks **alone**, of people of their scientific standing, that live in London, are excluded—for it is *exclusion* in such a case & remarked by every one to be so.

Ellen Busk was known as a freethinker and religious sceptic, and Charles Lyell had quarrelled with Richard Owen; the reason for Louisa and William Benjamin Carpenter's exclusion is not known. Hooker's remarks, whether his suspicions were accurate or not, suggest a way in which a scientific wife could exercise a kind of non-financial patronage, by controlling who met whom in the more relaxed surroundings of the family home. They also suggest that this informal role was perhaps beginning to seem unsuitable in the increasingly professional world of science.

Emma, unlike Mary Lyell, didn't aspire to hold regular soirées. Darwin found social events stressful, and part of the point of their seclusion in Down village was to limit the number of people he had to see. Nevertheless, as Darwin's fame increased, so did the number of visitors. Darwin's scientific status brought his family into contact with people they might not otherwise have met: the toll on their and Darwin's nerves was sometimes extreme. Emma described to William a visit on 6 September 1879 from Ernst Haeckel, an enthusiastic and polemical German Darwinist:

We have just had a visit from Häckel—v. pleasant but Oh—such shouting. it deafened F. [*father*] so you may imagine my feelings.

He has been coasting round the N. of Scotland & I suppose shouting against the winds & waves, & has not been able to let down his voice. He is employed about the collections of the Challenger— He is an active man, gets up at 5 or 5.30 in summer—breakfasts at 6 on a piece of bread & coffee & dines at 1. & is w. his family for some little time & then 3 or 4 more hours work—

He is like a great good-natured boy, & talks most devoutly about the way the German scientifics quarrel— & one can hardly imagine him being rancorous, w he certainly is—

Emma's main strategy for protecting Darwin's health was to force him to leave Down for a holiday or a visit to family. Visits to family in London could also be relaxing or taxing depending on how much company was threatened, and of what sort. Generally when the Darwins visited London, which they usually did twice a year for a week or more, they stayed with Darwin's brother, Erasmus. His literary friends were unlikely to get Darwin very worked up. After Henrietta married, her parents divided their time between her house and Erasmus's. Initially, in 1873, Henrietta offered to keep her own friends away for the duration of the visit, but Emma demurred:

I think we will fix on the 8 or 10th Nov. if that suits you, but we shall not agree to your tabooing all your friends—as they do not tire F. like seeing his own. I aim at his seeing nobody but the Huxleys & not giving luncheons at all—

We will stay a week—I should like to say 10 days—but I don't think I shall compass that.

Emma had expected her role as a nurse even before their marriage, writing on 30 December 1838:

> Maer
> Sunday

My dear Charles

I am rather ashamed of writing to you so soon again but if I disguise my writing in the direction I am in hopes the post master at Newcastle will think it is somebody else. I want to persuade you dear Charley to leave town at once & get some rest. You have looked so unwell for some time that I am afraid you will be laid up if you fight against it any longer. Do set off to Shrewsbury & get some doctering & then come here & be idle. I can't make up my mind to only a flying visit here & I shall be rather jealous if you stay longer at Shrewsbury than here & I am sure the D^r will think it highly contrary to the strictest etiquette so be a dear good boy. I am sure it must be very disagreeable & painful to you to feel so often cut off from the power of doing

your work & I want you to cast out of your mind all anxiety about me
on that point & to feel sure that nothing *could* make me so happy as
to feel that I could be of any use or comfort to my own dear Charles
when he is not well. If you knew how I long to be with you when you
are not well! You must not think that I expect a holiday husband to
be always making himself agreeable to me & if that is all the "worse"
that I shall have it will not be much for me to bear Whatever it may
for you. So don't be ill any more my dear Charley till I can be with
you to nurse you & save you from bothers.

The wives of scientific men often provided important professional assistance, whether as editors, translators, readers, or research assistants. Wid-
ows often took charge of the literary legacy of their husbands, collating
unpublished works and writing biographies. Marion Bell, for example, was
asked to organise the papers of her husband, the well-known surgeon and
author Charles Bell, after his death. He was the author of *Essays on the
anatomy and philosophy of expression*, which Darwin had cited approvingly in
Expression of the emotions in man and animals, published in 1872. Darwin evidently sent a presentation copy of his book to Marion, who sent her thanks
via her friend Fanny Wedgwood, the wife of Emma Darwin's brother Hensleigh, already encountered in chapters 2 and 3.

47 Albany Street N. W.
Monday 4th. Nov.^r—

My dear M^{rs}. Wedgwood—
I could say much to you of my thanks to M^r. Darwin, and of my
interest in his Book, and especially for the—*Restoration* of the old fash-
ioned Anat^y of Expression, with his gratifying words of the Author.
How I wish they had been acquainted!
The last book that C. B [*Charles Bell*] read, (leaning on the Bedroom
Chimneypiece at Hallow Park) on Thursday Ev^g 28th of April 1842,
was Darwins Voyage in the Beagle. and—he knew the love of Dogs
so well.—
...
His restorative Work, in the Portmanteau was *The Anat^y of Expression*, with the many notes, and additions, and emendations,—strewed
in, among the cut leaves of the old Edition.— At that sad period—
the *whole* got into confusion, and my B^r. Alexander said that *I* must
come to him to aid in what had been intended.
...
I am far behind in the race of today, with a deep respect for those
who love, and humbly search for truth. Do tell Mr Darwin of my
gratitude and Believe me, dear friend, | Yours affectionately Marion
Bell

In Marion Bell's life one can glimpse a kind of family industry; her brothers John and Alexander Shaw were also surgeons, and John lived with her and her husband until his death in 1827. After her husband's death she lived with her brother Alexander, and their house in London was a centre for scientific and literary society. Charles Bell's posthumous publications were edited by Alexander Shaw, presumably with help from Marion.

Darwin wrote formally to Emma in 1844 to ask her to publish his species theory if he died before he could publish it himself.

Down.
July 5ᵗʰ. —1844

My. Dear. Emma.

I have just finished my sketch of my species theory. If, as I believe that my theory is true & if it be accepted even by one competent judge, it will be a considerable step in science.

I therefore write this, in case of my sudden death, as my most solemn & last request, which I am sure you will consider the same as if legally entered in my will, that you will devote 400£ to its publication & further will yourself, or through Hensleigh [*Wedgwood*], take trouble in promoting it.— I wish that my sketch be given to some competent person, with this sum to induce him to take trouble in its improvement. & enlargement.— I give to him all my Books on Natural History, which are either scored or have references at end to the pages, begging him carefully to look over & consider such passages, as actually bearing or by possibility bearing on this subject.— I wish you to make a list of all such books, as some temptation to an Editor. I also request that you hand over him all those scraps roughly divided in eight or ten brown paper Portfolios:— The scraps with copied quotations from various works are those which may aid my Editor.— I also request that you (or some amanuensis) will aid in deciphering any of the scraps which the Editor may think possibly of use.— I leave to the Editor's judgment whether to interpolate these facts in the text, or as notes, or under appendices. As the looking over the references & scraps will be a long labour, & as the **correcting** & enlarging & altering my sketch will also take considerable time, I leave this sum of 400£ as some remuneration & any profits from the work.— I consider that for this the Editor is bound to get the sketch published either at a Publishers or his own risk. Many of the scraps in the Portfolios contains mere rude suggestions & early views now useless, & many of the facts will probably turn out as having no bearing on my theory.

With respect to Editors.— Mʳ Lyell would be the best if he would undertake it: I believe he wᵈ find the work pleasant & he wᵈ learn some facts new to him. As the Editor must be a geologist, as well as Naturalist. The next best Editor would be Professor Forbes of London. The next best (& quite best in many respects) would be Professor *Henslow*??. Dʳ Hooker would perhaps correct the Botanical Part

probably—he would do as Editor— Dr Hooker would be **very** good
The next, Mr Strickland.— If no⟨ne⟩ of these would undertake it, I
would request you to consult with Mr Lyell, or some other capable
man, for some Editor, a geologist & naturalist.

Should one other hundred Pounds, make the difference of procuring a good Editor, I request earnestly that you will raise 500£.

My remaining collection in Natural History, may be given to anyone or any Museum, where it w^d be accepted:—

My dear Wife | Yours affect | C. R. Darwin

If there sh^d be any difficulty in getting an editor who would go
thoroughily into the subject & think of the bearing of the passages
marked in the Books & copied out on scraps of Paper, then let my
sketch be published as it is, stating that it was done several years ago
& from memory, without consulting any works & with no intention
of publication in its present form—

PS | Lyell, especially with the aid of Hooker (& of any good zoo-
logical aid) would be best of all

Without an Editor will pledge himself to give up time to it, it would
be of no use paying such a sum.—

Fortunately Emma was not called upon to take up the burden, as *Origin of species* was published in 1859.

Acting as an potential literary executor was not Emma's only scholarly
role: she also read to Darwin, was his amanuensis, and translated French,
German, and Italian. According to a neighbour, Louisa Nash, she collected up references to his work in journals and newspapers. Her attitude
to the content of his work was at times brisk. In 1871, Darwin was wondering whether to try to have an article by one of Darwin's American supporters republished in Britain. 'I do not much care much about these things
& shall therefore be a good judge whether it is very dull', she announced,
according to Darwin (*Correspondence* vol. 19, letter to A. R. Wallace, 12 July
[1871]). As an amanuensis she learned to exercise restraint, commenting
to Henrietta in 1874: 'I percieve that F. dislikes, & is fidgetted nearly as
much by an alteration that is a manifest improvement as by a useless one.
When he dictates he dislikes so m. any sort of suggestion of improvement
that I have left off ever giving one— It seems to give him a jolt, & stops his
thoughts' (DAR 219.9: 114).

Emma was fully involved in the social life of Down village, but the only
fellow 'scientific wife' in or near the village was Ellen Lubbock, the wife
of Darwin's protégé John Lubbock. Darwin, according to his daughter
Henrietta, was fascinated by Ellen (DAR 246: 24); Emma thought she was
extravagant in her dress. John Lubbock's biographer wrote of her:

Ellen was an immense help to John in his early scientific days, she
acted quite as a secretary, kept his papers in order, looked up references, did all the diagrams for his lectures. She was an exceptionally

charming letter-writer, and kept up all the correspondence with his scientific friends. She wrote a beautiful hand and was a clever, brilliant woman, with the kindest heart and the most genial manner, and made all the friends who came to Lamas (and when the elder Sir John died it was all kept up at High Elms) so welcome and happy that everyone felt at ease with her. (Hutchinson 1914, 1: 79.)

She wrote for Francis Galton's *Vacation tourists* an account of her and her husband's archaeological researches in Denmark and travels elsewhere in northern Europe. Her chapter includes an account of the archaeological importance of the Danish Stone Age shell-mounds, or middens, a summary of the progress made so far in excavating them, and a first-hand description of them (p. 365: 'Anything so exciting as that day of grubbing among the oyster-shells I never experienced'). In 1875, she reviewed Darwin's *Insectivorous plants* for the *Academy*, a journal concerned with literature, philosophy, and science.

Ellen's letters to Charles and Emma frequently had to do with meetings with other men of science. The following letter, of 1 October 1866, is typical:

Dear M̲ʳ̲ Darwin

M̲ʳ̲ Herbert Spencer, who is staying here, has a very great desire to see you, and I write to ask if you would allow him to call upon you, or if you would, in the course of one of your rides, stop here for a short time—

We should be only too glad if we could persuade you either to lunch or dine with us—but please will you tell us what you feel equal to, or if you would rather not see M̲ʳ̲ Spencer at all.

With our kindest regards to you all, believe me yours most truly
Ellen Lubbock

Darwin also appreciated Ellen's observations on subjects of interest to him, although she deprecated her own skill in science, aligning herself with literature:

Dear M̲ʳ̲ Darwin

When you spoke of studying *expression*, I fancied you meant that of feature— It occurs to me that you may allude to forms of speech: but however, in case my first idea was right, I have turned down a page of "Adam Bede" which may interest you, if you had forgotten it— It is painful to feel that one can only be referred to as a student of works of fiction— but after a dose of Relationships, one requires a "halfpenny worth of bread to all that sack". And at any rate, the line I have taken in literature is thoroughly novel.

It was such a pleasure to see you & M̲ʳˢ̲. Darwin this morning. If you knew *how* great, I think you would come oftener.

Believe me yours most truly | Ellen Lubbock
Page 132. I have put a cross.

Ellen attended meetings of the British Association for the Advancement of Science with her husband (it was one of the few scientific meetings that women could attend), and involved herself energetically on behalf of the Ethnological Society, which merged with the largely anti-Darwinian Anthropological Society in 1871. The Ethnological Society, founded in 1843, had been an offshoot of the Aborigines' Protection Society, and to some extent maintained that society's liberal outlook. This letter is from November 1873.

Dear Mrs. Darwin
…
I hate begging—so now you will perceive I am going to beg. Yesterday I was at the Busks', & M⸳ Busk was groaning & lamenting over his Presidency of the Anthro—(I never *can* spell the horrid word)— Society—the name irritates him, as it does John, & it *isn't* the right one. We never wanted to be merged & swallowed whole in and by this mushroom society, with no good men in it— So I said well, why not alter it back to the Ethnological, which was the first & real root of the thing? To which he said despondently that they were in debt £700.

I said we would collect it: on which he brightened up & said if we could screw together *half* that, he should be in a position to say "take this if you become the Ethnological again: otherwise it will be returned to those who gave it."

Now I want M⸳ Darwin if he will to head the subscription: his name will go so far— I don't ask for the actual money, because we may never come to realize the sum wanted: but I ask for his name & a promise of something, if we see our way. John would help I know, but I don't like to speak for him. Poor old M⸳ Crawford would have given every penny he had, in fact I should think he turned in his grave when his pet Society was named after his bitterest enemies. Will you ask M⸳ Darwin to suggest a few people we might apply to.—

I am afraid I beg badly— I never did it but once before—but I am rather heart & soul in this matter. You see John was President for some years & the amalgamation or rather swamping of his society vexed him very much.

I hope you won't think me very meddlesome   I felt so sorry for M⸳ Busk— I don't think he's well, either.

With love to you all I am yours affectionately EFL

She also wrote verse about insectivorous plants after the publication of Darwin's book about them:

From the Insects to their friend, Charles Darwin.

We saw that you were watching us,
We felt you were our friend,
And as we, in a general way,
Come to a fearful end,

It suddenly occurred to us
That we would have a look
At what you wrote about us,
So we crawled upon your book.

We now have buzzed all over it,
And find that, as we feared,
Voracious Plants could tell us
How our friends have disappeared.

(**I** never trusted Drosera,
Since I went there with a friend,
And saw its horrid tentacles
Beginning all to bend.

I flew away, but *he* was caught,
I saw him squeezed quite flat—
I don't go any more to Plants
With habits such as that.)

We are very much obliged to you
For now of course we shan't
Be taken in and done for
By any clever Plant.

But *this* has to be considered:
It isn't much we need,
But if we daren't go to any plant,
On what are we to feed?

We feel that you, in pointing out
The dangers that we run,
Have meant to do the kindest thing
To us that could be done.

Therefore, to your abode in Down,
With joyful buzz and hum,
From every quarter of the globe,
We Insects all will come.

Great plates of honey you will set
For us upon your lawn,
We'll feast away & bless the day
That ever you were born!

E F Lubbock | 1875.

Frances Harriet Hooker came from a scientific family, being the daughter
of Darwin's Cambridge mentor, the professor of botany John Stevens Henslow. She married Joseph Dalton Hooker, Darwin's closest friend, in 1851.
In a letter to his grandfather Dawson Turner, Joseph commented on the
suitability of her background and connections, adding, 'She is much cleverer than I am & will I hope correct the press well ... The lady has plenty
of faults, but ... I do not see that I could do better.' (Quoted in Lightman
ed. 2004.) Although Joseph's approach to marriage seems far more cal-
culating than Darwin's, he became very much attached to Frances. She
assisted him significantly in his published work, and translated Le Maout
and Decaisne's *General system of botany* from French into English. Emma
disliked her, but to her relief, neither Frances nor her husband ever realised
it, as far as she could tell. Emma wrote to Henrietta in April 1873, after
Frances had made an extended visit, 'We were dismayed to find that Dr H
thought it likely she wd stay longer, so that I think & hope his eyes are not
opened to her utter want of pleasantness' (DAR 219.9: 101).

Frances's letters to Darwin were usually written on Joseph's behalf when
he was too ill to write: this one was written in 1865.

My dear Mʳ Darwin
   Joseph made no progress to recovery at Kew—& so by Dʳ Quain's
advice, we removed to Notting Hill on Satʸ, where we are staying
with the Campbells—Joseph bore the little journey very well, & was
better on Sunday—
   Monday was a bad day, with much fever, which pulled him down
very much—but he is now better again, & says he is certainly gaining
strength— But he is very much reduced, & very unequal to any, even
the slightest exertion—
   Buxton will probably be the next move, when he can bear the journey—
   He asks if you saw the article of Mʳ Croll in the last Reader [*a journal*] on the displacement of the Earth's axis(?) by the ice of the glacial
epoch?— & also do you know that the Reader has been sold to the
Anthropologicals [*the Anthropological Society, mentioned above*]?
   Sir C Lyell told him this,—if true, Joe thinks it a great disgrace to
all parties concerned—
   Sir Charles was looking very well, & Joseph was delighted to see
him, though he unfortunately had a very severe fever attack at the
time—

We wish much to know how you are— Will you kindly let the Lubbocks know about this account of Joseph? as I am overwhelmed with letter writing—

Joe's love & mine to M^rs. Darwin, Etty & Lizzy & yourself—

Believe me | Your's aff^tly. | F H Hooker

Sep. 6.—

Like Ellen Lubbock, Frances Hooker accompanied her husband to BAAS meetings: he marvelled at her stamina. In this letter, written about a fortnight after the last, she mentions a railway accident in which Ellen, who was heavily pregnant, was injured.

> Buxton—
> Sep! 22—

My dear M^r. Darwin

Many thanks for your letter, which I waited to answer until we should be here— I can now tell you that we accomplished the journey very successfully in two days, passing the first night at Derby, & arriving here yesterday— We are at present located in a boarding house, but we do not like it, & are going out this morning to look for lodgings— The noise & publicity does not suit Joseph, & he was very tired last night— However he is really wonderfully better, & beginning to walk about comfortably—but he is still very stiff in his joints, & can only move slowly— It is horribly cold here, but I suppose we shall get used to that in time—but it is a great contrast to Notting Hill—

D^r. Tyndall, who was at Birmingham, told us that so far from Lady Lubbock having kept her room, she was sitting in the sections [*i.e., attending lectures*] all Monday!— the accident having happened on Saturday—

Joseph has read Phillips' address, & thought it in matter extremely washy, but no doubt as delivered, was very well suited to the occasion— He was astonished at no allusion whatever being made to Sir J. Lubbock's book.

With Joseph's love, | I am | Yours affect^ly. | F H Hooker

Have you read Geikie's book on Scotland? Joseph was pleased with parts, but disappointed on the whole.

In 1873, the scientific wives may have come together to raise money for T. H. Huxley, who was overworked and hard pressed for money. In Henrietta Litchfield's version, the initiative began with Katherine Murray Lyell, Mary Lyell's sister (they had married brothers).

> Mrs Lyell, in a talk to my mother, during this stay in London, suggested whether a very few of his most intimate friends might not

quite privately join in making a gift to enable him to get away. My father took eagerly to the idea, and became the active promoter of the scheme. (*Emma Darwin* (1904) 2: 262)

From Emma's diary, we know that this meeting between her and Mrs Lyell took place on 4 April 1873. On 6 April, Darwin wrote to J. D. Hooker to suggest raising a fund for Huxley. Around the same time, he also spoke to Ellen Lubbock, who had evidently been asked for advice on the delicate subject of how to break the news to Huxley.

Dear Mʳ Darwin

I have been thinking much of what we spoke of: and it occurs to me that if John either has not time, or does not feel sufficiently intimate with our friend to propose anything, Mʳ Hirst would do it gently, & with delicate tact. Or Mʳ Spottiswoode: but I think Mʳ Hirst is a nearer friend.

Of course, I can answer for John in other ways, & will see that he writes to you tomorrow: but I *do* think that when it is suggested, it ought to be quietly placed before our friend that not only for his own sake, or for that of his family, but for the sake of Science, he should be above the pride which might lead him to reject what we should feel it a privilege to offer.

With thanks to you for coming, I am yʳˢ most truly EFL

Hooker meanwhile responded enthusiastically on 7 April:

Fanny called on you the other day with some such a proposal on the tip of her tongue. She had suggested to me the paying to Huxleys bankers the amount of his law expences (to be raised by you, I, & a couple or so more)— I asked Tyndall the night before F. called on you, & he thought the affair too small, & that H. would not like the "stealing a march upon him." so we agreed that F. should say nothing to you about it. The matter has however never left our minds—

Frances's call on the Darwins may have been as long ago as 28 March. Did she really not say anything even to Emma at the time? Had she discussed the idea previously with Katherine Lyell, or Ellen Lubbock? Emma carefully noted, 'Meeting about Huxley' in her diary on 8 April: this was presumably a reference to her husband's meeting with Hooker. At all events, the campaign was from then on an exclusively male affair, and John Tyndall, one of the contributors, wrote to Darwin: 'I wish with you that Mʳˢ L. [*Katherine Lyell*] had not subscribed— It suggests the idea of *an effort*, which ought to be entirely absent from the movement' (*Correspondence* vol. 21, letter from John Tyndall, 16 April 1873). Mrs Lyell's name was deleted from the list of subscribers that was circulated to the eighteen contributors. Was it felt that she simply didn't know Huxley well enough? Or was it inappropriate for a woman to be helping a man financially, rather than urging other men to do so?

Raising money, applying for pensions, and getting jobs for other people is not an uncommon theme in letters to Darwin from women, as we have seen in the letters of Mary Butler and Ellen Lubbock. Arabella Buckley is another example. Buckley was Charles Lyell's secretary for more than ten years, and wrote and lectured on science (for more on her writing, see chapter 10). She was a close friend of Alfred Russel Wallace, the author, with Darwin, of the joint paper of 1858 that announced the theory of natural selection to the scientific world.

1 St Mary's Terrace | Paddington W.
Dec 16./79

Private

Dear Mʳ Darwin,

I want very much to consult you upon a matter in which I have perhaps no real concern, but with which I believe I am better acquainted than others—

You will no doubt have known that Mʳ Wallace was a candidate for the post of Superintendent of Epping Forest & has been making great efforts to get it during the whole past year. He is now rejected & they have chosen a landscape-gardener instead—

Now he is so modest & sensitive about himself that I am sure he would never tell anyone that which however I know, that "pecuniarily it was of importance to him to get a regular salary".

He is not strong & literary work tries him very much & the uncertainty of it is a great anxiety to him—

In a letter to me the other day he writes "I want some regular work either partially outdoor, or if indoor then not more than 5 or 6 hours a day & capable of being partially done at home— This I see no probability—hardly a possibility of getting at my age & with my irregular antecedents"—

Now I cannot help thinking that if men like Sir J. Lubbock, Sir J. Hooker & others knew that Mʳ Wallace wanted work of a modest kind & not some important post, some good use might be made of his great Natural History power & his future made more secure— Only, of course, my moving in the matter should not appear, I merely suggest that, which if it could come, must do so from men of his own standing & I shall not mention to any one that I have written to you— Years ago he was to have had the East London Museum but it passed into the hands of S. Kensington & he lost it— I feel he *ought* to have something & I could think of no one as good as yourself to whom I could say so—

I remain | Yours very sincerely Arabella B Buckley

Darwin's response was initially positive, but after consulting J. D. Hooker, he decided that Wallace's interest in spiritualism (an interest Buckley shared with him), and some other instances of apparent poor judgment, would make it impossible to ask for support for a petition for a government pension for Wallace. Buckley responded: 'I suppose it is hopeless, & indeed I have always feared that Mr. Wallace's want of worldly caution might injure him, though he would be a most valuable man in the right place' (letter from A. B. Buckley, 20 December 1879, DAR 160: 368).

However, in 1880, Darwin decided to raise the subject again. This time he wrote to Buckley:

> 4 Bryanston St. Portman Sq.
> Sunday Oct. 31.

Home tomorrow or early on Tuesday
Private
My dear Miss Buckley

Some time ago I spoke to Sir J. Lubbock about Wallace and a Government pension, and this morning I produced a decided effect on Huxley.— He has asked me to draw up a full, but *condensed* statement of Wallace's claims; and he will then endeavour to talk over Hooker and Spottiswoode.— Therefore I think there is a fair *chance* of getting up a memorial to Government.— When I began to think over the case, I found myself very deficient in knowledge, and bethought me that you with your generous spirit would aid me. I have written down some questions, which will serve me as memoranda when I get home, and when I will lose no time.— If I were to ask Wallace any of these questions he would think me mad or impertinent.— (He perhaps would think that you intended writing sketch of his life in some Journal.) Perhaps you can answer some, or get answers by some indirect manner from him.— Any hints or advice of any kind would be of greatest value.— Especially about his present circumstances. You will understand these materials are solely for Huxley, Hooker and perhaps 2 or 3 others' consideration.— The Government Memorial will be a separate consideration. I do *most earnestly* hope that we may succeed.

I know well, busy as you are, that you will help me as far as lies in your power.

Believe me, My dear Miss Buckley | Yours sincerely | Charles Darwin

Huxley feared that even if we could get a memorial signed by a few first-rate men, yet it might be extremely difficult to get a pension on account of the scandalous manner in which these pensions are jobbed.—

Therefore it seems very desirable that Wallace should hear nothing about it

1 St Mary's Terrace | Paddington W.
Nov 7. 1880.

Dear M.ʳ Darwin,

It is all right as far as regards M.ʳ Wallace himself—

I told him that you & M.ʳ Huxley thought him entitled to a gov.ᵗ. pension *if* it could be got— At first he hesitated but when I represented that such men as Joule & Faraday had received it he said "I confess it would be a *very great relief* to me and if such men as Darwin & Huxley think I may accept it it suppose I may"—adding "I really have some claim, for most naturalists & travellers on their return from a foreign country have been given some post, & I have tried for one in vain".

It seems some friend suggested it to him some time ago but he rejected the idea; but now that it comes from men like yourself & Huxley who can appreciate his work it makes a difference—

I could not get the memorial lists but when I said that you would have only a few good names & suggested the Duke of Argyll, M.ʳ Wallace said he is just the man who would probably give his name with pleasure—

He quite understands that the result is very doubtful & indeed he said very little about it, for when I had once ascertained his views I did not want to lead him to dwell upon it—

I have nothing I think to add to the notes I gave you. I enclose a very brief statement which may be of some use, though of the real value of his work you can speak best—

F⟨ro⟩m my short conversation yesterday I am more than ever sure that your generous efforts if they succeed ⟨wi⟩ll really confer a great boon on M.ʳ Wallace & relieve him of anxiety—

If I can look out anything more for you please let me know—

With kind remembrances to Mrs. Darwin | Yours very sincerely
Arabella B Buckley

The appeal was successful.

Villa Margherita San Remo—
Jan 13. 1881

Dear M.ʳ Darwin

I *am* glad & congratulate you most heartily on the success of your generous undertaking

Of course you were the only person who could tell M.ʳ Wallace— He may well be proud both of his proposer & seconders, & it is this which will make the pension a pleasure as well as a boon to him.

Beyond being proud of the instinct which led me to state the particulars to *you* as the right man, I have really had nothing to do with it—

I have always felt that your generous friendship for M.ʳ Wallace, & the almost overdue credit which you have always assigned to him, is one of those bright spots in the history of science, which ought to shame all those who indulge in petty jealousies; & this success is the befitting crown to the whole matter—

I shall now write & congratulate him, telling him that you have let me know, though of course it must not be spoken of till officially announced— I will also let him know who are his supporters.

I am so glad it has come just now when he is looking forward to settling down in a month or two in his little cottage at Godalming, so that he can work with his garden & his writing without feeling the pressure which has forced him of late to work at uncongenial writing—

With most sincere congratulations | I remain | Yours very sincerely | Arabella B Buckley.

Women's work assisting scientific men was probably most uncontroversial when they worked as editors and secretaries. Their role as scientific hostesses and advocates was sometimes contentious, and possibly became more so as science became more institutionalised. It was a more obvious way of wielding power rather than simply being of assistance, and by the second half of the century may have been going out of fashion. Their role as fund-raisers was important but deliberately muted; their role as funders decidedly awkward. However, for many wives, a husband's scientific career was a joint endeavour.

5   Observing plants

Botany was a popular subject for women in the nineteenth century. The materials were readily accessible for home study, and it was thought to be a good way of encouraging women to go outside and get some exercise and fresh air. It was, furthermore, an important subject; medical students studied botany as an essential part of their syllabus (materia medica, the raw material of medicines), and the increasing importance of empire, together with new experimental approaches such as Darwin's, rendered it cutting-edge. Of more than six hundred letters exchanged by Darwin and female correspondents, the largest number, after letters about family matters, are about botany. These range all the way from observations carried out on his behalf by nieces, to exchanges with other specialist botanists. This chapter only has room for a fraction of the letters available, and concentrates on four correspondents: Dorothy Nevill, Lydia Becker, Mary Treat, and Sophie Bledsoe Herrick.

Lady Dorothy Nevill had a notable garden at Dangstein near Midhurst, Sussex. She specialised in the cultivation of orchids, pitcher-plants, and other tropical plants. Her head gardener was James Vair, and since in her letters and autobiography she always minimises her own knowledge and accomplishments, it is difficult to assess how expert she was herself. Joseph Hooker wrote, 'She was not the frivolous character she paints. She was thoroughly interested in the rare plants of her noble garden' (Nevill 1919, p. 66). She was a tireless correspondent, supplied Darwin with many plants, and read his books. Darwin first wrote to her in 1861.

> *Down. | Bromley. | Kent. S.E.*
> Nov. 12th

Madam

D[r.] Lindley has told me that he thought that your Ladyship would be willing, if in your power, to assist me.— I am preparing for publication a small work "on the various contrivances by which Orchids are fertilised." I much wish to examine a few more exotic forms, & if you happen to have those which I wish to see, possibly your Ladyship would be so generous as to send me two or three flowers. I am aware that it would be a remote chance that your Ladyship should possess or spare these flowers. I chiefly want any member of the great Tribe of Arethuseæ, which includes the Limodoridæ, Vanillidæ &c.

Lady Dorothy Nevill
© National Portrait Gallery, London

Mormodes & Cycnoches are especial desiderata, though they would be most difficult to send, as the pollen-masses move or explode when the end of the column is touched. I, also, want much Bonatea, Masdevallia & any Bolbophyllum with its lower lip or Labellum irritable. Indeed any genus with any remarkable peculiarity would be most gratefully received.—

I have much reason to apologise for thus intruding on your Ladyship; & I am far from expecting that your Ladyship will reply to this note, if you cannot assist me.— I will only add that for a parcel too large for the post, my quickest address is "C. Darwin care of the Down Postman Bromley Kent". I find that orchids travel safest in tin boxes or cannisters with a little damp paper.—

With many apologies, I have the honour to remain | Your Ladyships | Obedient servant | Charles Darwin

This letter from Nevill was written before 22 January 1862.

Dangstein | Petersfield

My dear Sir

I am most grateful for your most kind letter and the promise of the Photo. already I am making a place for it amongst my other friends It has not arrived yet but I must be patient for I know in this dull weather they cannot make copies

I grieve to say I have at present no melastomaceous plant in flower— We had a glorious Pleroma but it is over— My gardener says that he has observed that when the flower of the Cynoches is quite dry by the least touch the anther appears to go quite back with a jerk and at the same time it ejects its pollen which it throws on to the pistil   I do not know whether I have described it rightly or not I am most grateful for your little pamphlet [*probably one of Darwin's botanical papers*] We have had a house full but now that I am alone I mean to give great attention to it

Thanking you again many times | believe me | most truly yours | Dorothy

The correspondence lapsed in 1862 with the publication of *Orchids*, and was resumed in 1874.

Dangstein Petersfield
2nd Sep^{ber}

My dear Sir

Some years ago you were kind enough to refer to me on some question relating to the fructification of Orchids— I have been most deeply interested in your last investigation into the carnivorous properties of certain plants and Dr Hooker has told me that you would much like a plant of Dionea Alas ours are all too small to be of any use to you but we have plants of Drosera Dichotoma and we could send you pitchers of the different kinds of Sarracenia— I am sure we possess numerous plants which would interest you could you but come and see them— need I say what pleasure it would give both Mʳ Nevill and myself were you to do so— I know you are interested in Cats I have got a beautiful one from Siam— I fear we can never tempt you here but could I at any time be of use to you in sending plants etc I shall be delighted to do so— so pray make use of us— I do so regret that at present the Dionias are not of any use— Do you know the Pulmonaria Maritima or Oyster plant—

believe me | most truly yours | Dorothy Nevill

Down, | Beckenham, Kent.
Sep 7. 1874

Dear Lady Dorothy Nevill

The Drosera arrived quite safely on Saturday night. I fear it did not like its journey, as the glands are rather dry; but I hope in a few days to see them secreting, & I will then make my observations. I have put the plant in a cool hothouse, which I trust is right. I will look in the course of the day & try to discover whether the pot has stood in a saucer of water.

As you were so kind as to offer to aid me, I will mention a plant which it is possible, though very unlikely, you may possess, namely an Epiphytic Utricularia. This plant, when making fresh shoots or leaves, produces minute bladders, but at no other time; & I am most anxious to examine a few of these bladders.

The Drosera is an extraordinary looking plants & I am grateful to you for the opportunity of observing it.

Your Ladyship's | truly obliged | Charles Darwin

Dangstein | Petersfield

My dear Sir

We keep the Drosera in a house with a north aspect and hardly any heat—and damp— The gardener says you must syringe it and that will produce the dew on it—in a short time— it does not stand in a saucer of water— We have small plants of Utricularia Montana which may perhaps be the plant you mean— They are seedlings from our parent plant and we could well spare you one for good

I can assure you it is a great pleasure as well as an honour to contribute in however small a degree towards your interesting investigations— I only wish (when I am in London) that I might have the pleasure of coming down for an hour and making your personal acquaintance but perhaps that would be too much to expect but always believe anything we have we will willingly send you

believe me | Y^s truly | D Nevill

8th.

Dangstein, Petersfield
16th

My dear Sir

I am so pleased and gratified by your photo received safely this morning— We looked particularly at the U— we sent you and if you remove the moss you will find the bladders at the footstalks— which I think are what you want   The Gardener shewed them me before sending off the plant to you so I do hope you will find what you require— We did not in the least require the Drosera to be returned— We have put on 4 leaves of the D Capensis bits of meat the leaves immediately curled up and after 4 days the bits of meat disappeared— we placed earth and moss on other leaves but there was no effect whatsoever We placed bits of white grissle in a pitcher of Nepenthes Rafflesiana—and another kind—and they have both been materially lessened but more so in the pitcher that contained the most fluid—

If you would only pay us a visit here you would find an immense field for experiments and it would give us the greatest pleasure to see you— Do you know anything of the enclosed vegetable snails— and worms— They are from America—and perhaps might interest you   I fear there is no chance of our getting you down here but pray remember that if at any time we can help you with specimens we shall be truly glad to do so   We leave this on the 28th for about 10 days

believe me | Y^s obliged | D Nevill

Down, | *Beckenham, Kent.*
Sept. 18th

Dear Lady Dorothy Nevill

I am so much obliged to you. I was so convinced that the bladders were with the leaves, that I never thought of removing the moss, & this was very stupid of me. The great solid bladder-like swellings almost on the surface are wonderful objects, but are not the true bladders. These I found on the roots near the surface & down to a depth of 2 inches in the sand. They are as transparent as glass,—from $^1/_{20}$th to $^1/_{100}$ of inch in size, & hollow. They have all the important points of structure of the bladders of the floating English species, & I felt confident I sh^d. find captured prey. And so I have to my delight in two bladders with clear proof that they had absorbed food from the decaying moss. For Utricularia is a carrion-feeder & not strictly carnivorous like Drosera &c &c.

The great solid bladder-like bodies, I believe are reservoirs of water like a camel's stomach. As soon as I have made a few more observations, I mean to be so cruel as to give your plant no water & observe whether the great bladders shrink & contain air instead of water. I shall then, also, wash all earth from all roots & see whether there are true bladders for capturing subterranean insects down to the very bottom of the pot. Now shall you think me very greedy if I say that supposing the species is very precious & you have several, will you give me one more plant, & if so please to send it to "Orpington St^n S.E.R^y to be forwarded by foot-messenger".—

I have hardly ever enjoyed a day more in my life than this day's work; & this I owe to your Ladyships great kindness.

The seeds are very curious monsters: I fancy of some plant allied to medicago; but I will show them to Dr Hooker.

Your Ladyship | Very gratefully | Ch. Darwin

Dangstein, Petersfield

My dear Sir

I was so pleased to receive your most interesting letter and its contents I am only too delighted to contribute in any way to your studies The Gardener has just brought to me 2 Utricularias we have— One is our specimen plant just coming into blossom and the other is one smaller than the one sent to you and he fears too small to be of any use to you at present so had we not better keep it for a time till it has developed its useful parts Dont scruple to take it as we have 8 more seedlings but I daresay in a month or so it will grow enough for your purpose— We have been placing bits of meat on the Drosera Capense and it is curious to observe how its leaves closes over the meat and when quite consumed how they relax to their ordinary state

ever most truly yours | D Nevill

22

Lydia Becker was a Lancashire woman, the daughter of an industrialist, and was educated mostly at home. While living in country, she developed an interest in botany. She later became famous as a leader in the suffrage movement, and as an advocate of women in science (see chapter 14). She wrote her first letter to Darwin very politely in the third person about some hermaphrodite flowers of *Lychnis dioica* or *diurna* (red campion, now *Silene dioica*). Usually, the flowers of red campion are either female, i.e. they have pistils, or male, i.e. they have stamens, with anthers.

Miss Becker presents her compliments to Mr Darwin and takes the liberty of sending him the enclosed flowers of a variety of Lychnis dioica common in the woods here but which she has not observed elsewhere. It has bisexual flowers and large dark purple anthers which give the plant a very striking handsome appearance. The same conspicuous anthers occur in flowers bearing stamens only of which one is sent. Miss Becker does not know whether such a variety as this would be interesting to M.ʳ Darwin, if not she must apologise for having troubled him.

Altham | Accrington | Lancashire
May 18.ᵗʰ 1863.

Evidently Darwin was interested (his letters to Becker do not survive), for Becker wrote again.

> Altham | Accrington.
> May 21

Sir

I have this day forwarded to you a small box containing plants of Lychnis which I hope will reach you tomorrow in tolerably good condition. I have also enclosed a tin canister of flowers and I hope there will be pollen enough to enable you if you please to try the experiment of fertilising with it some of the plants grown in your neighbourhood. Since I wrote last I took my note book into a different part of the wood and examined 122 plants of Lychnis with ⟨the⟩ following result—

Common form. males 41 females 23. hermaphrodite 29 males with dark stamens. 29 I pulled to pieces numbers of the flowers with long p⟨is⟩til but never found the stamens developed in them. The stamens are normally of medium length but in some plants they are very long, in others very short, the flowers on each plant are generally alike in this as in other respects.

With many apologies for troubling you by writing so much about these plants I remain Sir | Yours very respectfully | L. E. Becker

> Altham | Accrington.
> May 23.ʳᵈ

Sir—

Allow me to thank you for your most kind and courteous reply to the communic⟨ation⟩ I ventured to make to you I am indeed grateful that you wish to investigate the structure and history of the curious Lychnis. I hope in the course of next week to send you a small hamper containing roots which will probably continue to flower this season after they have been planted in your garden. On receipt of your second letter this morning I went into the wood and examined 137 plants of Lychnis, taking them indiscriminately, though I do not pretend to have examined all I saw. The plants were of four kinds, viz

1 Male flowers of the usual type small pale yellow anthers   56
2 Female flower do. with long spreading pistils   25
3 Hermaphrodites having large dark purple anthers
  and short straight upright p⟨istils⟩   31
4 Male flowers with stamens like those of the
  hermaphrodite.   25

I searched in vain for a female flower with pistils like the hermaphrodite or an hermaphrodite with pistils like the female flower. this suggests the query Can the long spreading pistils be an adaptation or a struggle to catch the pollen wafted from a distant flower, a provision needless in the hermaphrodite which has an abundant supply close at hand.

I tried in vain to classify the plants according to the ⟨length⟩ of the stamens   this ⟨seems to⟩ vary indefinitely even in the same flower—And I think that when the plant has been gathered a few days, the stamens do grow rather long; you will readily sort the flowers I sen⟨t⟩ under the four heads I have enumerated. I have put in one or two of the common type for comparison with those which grow in your district. I have been given to understand that hermaphrodites occasionally appear in this Lychnis, but have never been able to make out whether they normally differ so much in the character of their stamens and pistils from the common form as do those found at Altham. I have never seen an hermaphrodite Lychnis diurna except in the woods ⟨near⟩ here where it is, ⟨as⟩ you may gather from the numbers given above, very abundant, and so I have observed it several years in succession, I fancy it is hereditary. I will take care to collect as much se⟨ed⟩ as I can during the summer marking and noting the plants from which I obtain it, and transmit it to you at the end of the season in the capsules. I will also find out as much as possible of the distribution numbers and range of the variety. There is a practically unlimited supply of specimens here and I need hardly say that I shall be most happy to send you as many, and whenever you would like to have them. I hope it is not asking too much that y⟨ou⟩ will be good enough to favour me ⟨wi⟩th the result of your investigation and also let me know if you hear of the variety occurring in other places.

I am Sir | yours very respectfully | Lydia E. Becker

May. 24. I have kept the flowers fresh till today to avoid the detention in the post office over the Sunday. I hope they will arrive in good condition but I will enclose a tin can also with a bunch of the flowers in the packet of roots.

Darwin evidently concluded that the stamens were infected by a fungus, and the destruction of the pollen had caused the pistil to develop in compensation.

Altham | Accrington.
May 28

Sir—

At the risk of being troublesome I cannot refrain from expressing my thanks for your letter and the paper you have done me the
honour to send me [*one of Darwin's botanical papers*] which I do indeed
highly prize. I had seen an abstract of it, and this induced me to send
you the Lychnis. I am of course much disappointed that it does not
possess the interest you at first thought it might possibly have but I
cannot regret having sent it as it has procured for me the pleasure
and honour of the communications with which you have favoured
me. The resemblance of the black powder on the anthers to the smut
of wheat had struck me but I never thought of the obvious inference
that it was a similar cryptogamic growth. And there are one or two
points which still excite my curiosity   How is it that this parasite
invariably attacks the hermaphrodites, and how does it get into the
unopened buds? Can it be that the plant is weakened in the endeavour to perfect both stamens and pistils and so rendered liable to the
attacks of disease. And does this presumed weakness in the plant
account for the small size of the pistils?

I found one solitary f⟨emale⟩ plant with small p⟨istils⟩ and upon
watching it for three days to see if the pistils would grow, I cut off
and send you the only flower expanded. Though the plant looks vigorous enough I suspect it is accidental weakness and not divergence
of form which causes the peculiarity but I will try to save seed from it
and also test the goodness of the seed of the hermaphrodites.

I am sorry to have troubled you with sending the roots now that
they are not likely to prove interesting   indeed I hope they will not
do mischief by spreading among other plants the insinuating parasite
with which they ⟨ar⟩e charged, and I beg you to believe that I am not
so unreasonable as to expect you to give yourself any further trouble
on my account though I should be very glad if my difficulties could
be satisfactorily explained.

Again thanking you for your extreme courtesy and kindness—I am
Sir—yours much obliged and very respectfully | L. E. Becker

Altham | Accrington
July 8th.

Dear Sir

I send you some seed of Lychnis diurna gathered in Altham woods.
You will probably not be surprised to learn that I have hitherto been
unable to obtain a single capsule of seed from an hermaphrodite
flower. I have examined, I should think, hundreds of plants, and
find, that invariably, the oldest capsules, corresponding in age to
those which in the female plants, have already ripened their seed, are
shrivelled up to nothing. Many of the younger capsules seem so fresh

and healthy, (though none are so vigorous as on the female plants,) that I cannot help fancying they may come to something, but I have never discovered a good one at all approaching maturity. I will continue to watch the plants carefully, and should I obtain seed from the hermaphrodites I will not fail to send you some.

I was prevented by wet weather from going to the woods as soon as I wished to mark the plants, and those I at length ticketed have scarcely ripened their seed, but I entertain no shadow of a doubt, from the appearance of the capsules and the remains of the styles, that all the seed sent is from female plants. It was gathered in a wood abounding in hermaphrodites which seem to be nearly if not altogether barren an⟨d⟩ it may possibly interest you to try whether they will produce hermaphrodites when sown in another locality. I send it now, as I have an impression that in order to have flowers next spring, it is advisable to sow it so that they young plants may get well grown before winter. I observe a considerable variation in the form of the capsules, these are usually nearly globular, but some plants have them shorter others longer than the average. The elongation is occasionally carried so far as to seem like an imitation of the allied genus Cerastium   I have put in separate papers a few capsules shewing the extreme forms, and I intend to try whether the peculiarities are hereditary, as well as constant in the individual, and can be perpetuated or increased by selection.

In your last letter you say that the pale colour of some of the flowers leads you to suppose that the white Lychnis may grow in the same wood, and that the hermaphrodites may be natural hybrids. But in Professor Babington: "Manual" [*Charles Cardale Babington, Manual of British botany*] it is stated that both Lychnis diurna and vespertina vary in colour from red to white and from white to red. I could have gathered a bunch in our woods, shaded through every tint from rich rose to pure white, which would have admirably illustrated this remark as regards diurna, but the whitest of them all did not make the slightest approach, in any other character than colour, to the genuine White Lychnis, which I have never seen growing in this district. And whenever I have been fortunate enough to meet with this very interesting and rather scarce plant, it did not grow in woods, but in open fields or hedges. Since I received your letter I have looked more particularly for it, but in vain.

It therefore seems as if we must look to another cause for the hermaphrodites, and there seems nothing but to attribute it to the parasite. You say, adding the perfectly useless caution not to trust you on the point that the fungus may exist in the tissues of the plant and be propagated in the seed   I could not, if I tried, help believing that this is so though I am at a loss to comprehend how the propagation is accomplished   The female plants appear thoroughly free from the parasite while the infected plants seem incapable of producing seed, and their pollen you have found to be entirely destroyed by the fungus. Is it possible that the spores of the fungus may fall on the stigma and be conveyed along with the healthy pollen tube to the ovules? They must get into the tissues of the plant in some way as the fungus is found in the immature, unopened buds and I cannot do⟨ubt⟩ the correctness of your suspicion that it destroys the pollen and causes the development of the ovarium in some, though not in nearly all of the infected plants—

I hope you will be indulgent with me for presuming to write you so long a letter and believe me to be | yours most respectfully | L. E. Becker

Becker reported her observations on *Lychnis diurna*, her correspondence with Darwin, and her own conclusion that a parasitic fungus had induced hermaphroditism in the flowers at the 1869 meeting of the British Association for the Advancement of Science (the discussion was reported in the *Journal of Botany*; see Becker 1869b). John Hutton Balfour, the professor of botany at Edinburgh, while acknowledging the excellence of the paper, disagreed with Becker's conclusion: if Becker was right, the instance was the first known to the botanical world. If it was, replied Becker, why might not she make the first discovery of it?

28 Jackson's Row | Albert Square | Manchester
Oᶜᵗ. 14. 1869

My dear Mᵣ Darwin

Will you permit me to send you the enclosed abstract of a paper read at the Exeter Meeting of the British Association, respecting the curious variety of campion flowers I took the liberty of sending you a few years ago, and in which you were good enough to express some interest

If you think the subject worthy of your attention I should be greatly obliged if you would kindly inform me whether you think it possible that the fungus could exert what looks like an "active coercive force to bend the structure of the flower to its necessities". From all that I have observed I cannot rest in any other conclusion but in the enclosed abstract, I have merely given the result to which I have been led and not entered into the detailed reasons for it.

If you paid any attention to those plants which I sent you, you may have observed some facts which will bear on this curious question.

On your theory of Pangenesis it seems to me not inconceivable that gemmules of stamens may circulate undeveloped in the pistilliferous plants of campion, and that the presence of the parasite may cause the condition under which these can develop—but without this theory I am totally unable to conjecture in what manner the fungus possibly can cause stamens to grow in a flower that would not naturally have produced them.

Apologising for troubling you with this note I am | yours very truly | Lydia E. Becker

Charles Darwin Esq—

⟨On alteration⟩ in the structure of Lychnis diurna ⟨observed in co⟩nnection with the development of a ⟨parasitic⟩ fungus

Abstract of Paper read before section D— British Association, Exeter—

Specimens were produced of the common red campion, *Lychnis diurna*, infested with a parasitic fungus allied to the "smut" in wheat which fungus develops its fructification in the anthers of the flower. The campion in its ordinary healthy state has flowers bearing stamens only or pistils only, but about half the plants infested with the parasitic fungus bear flowers containing both stamens and pistils. The writer had never observed bisexual flowers on healthy plants, and attributed the occurrence of that condition in the specimens produced to the action of the parasitic fungus. The diseased plants very rarely produce seed but occasionally late in the season perfect capsules bearing good seed are found on them. A few of the flowers had been submitted to M: Darwin, and he had suggested that the pollen being destroyed by the parasite at an early period, the pistil was developed in compensation. But the writer thought that the influence exerted by the parasite was of a much more subtle and surprising character ⟨than this,⟩ and that instead of causing the develop⟨ment of⟩ a pistil in compensation for destroyed ⟨stamens the⟩ fungus has the power to cause a plant which would naturally have produced pistils only to develope stamens for the accommodation of the parasite— She supposed that the spores of the fungus fell on the stigma of a healthy flower, and infected all the seeds produced by that capsule that of the plants thus raised all which were naturally male plants remain unaffected in structure by the parasite, but they have their pollen destroyed by it— that all which were naturally female plants would develope their pistils to a certain extent, but as the fungus cannot produce spores without anthers to fructify in or pollen to feed on, it compels the campion to develope these for its accommodation, and the effort of so doing exhausts the forces of the plant, and causes the decay of the capsule, if indeed the previous stunting of the styles does not prevent fertilization. The production of healthy capsules late in the season may be owing to decay in the vigour of the fungus when, the pressure being removed, the plant resumes its natural functions. The fact that only about half the diseased plants are bisexual favours the view that the latter are female plants in which the growth of stamens has been induced by the presence of the fungus—

Mary Treat was an American botanist and entomologist. She wrote many popular and scientific works from 1869 onwards. After she separated from her husband, she supported herself by her writing and by collecting insect and plant specimens. She began corresponding with Darwin about insects in 1871 (see chapter 7). Her first letter focusing on botany came in 1872, when she wrote about sundews (*Drosera*).

> Vineland, New Jersey,
> Dec. 13, 1872.

Mr. Darwin:
Dear Sir,

Prof. [*Asa*] Gray writes me that you have found the *nerves* in *Dionæa*. Good! And he asks me, in connection with himself, to make observations on *Drosera filiformis*, which I will gladly do.

As far as my observations extend, I do not consider this species so interesting as *D. longifolia*, or *D. rotundifolia*, although fully as *carnivorous* as the two latter, yet it captures only *small* insects which do not require any movement of the leaves to help confine them.

For some reason my plants did not work so well last season as the year before. Whether they were weakened by the unusually dry spring, or whether the locality from which I obtained them was not so good, or whether the fault may not have been somewhat with myself, I cannot say. …

Your theory is steadily gaining ground among the masses and thinking people of this country, Prof. Agassiz to the contrary notwithstanding. It is boldly advocated from an Orthodox pulpit in this place, and from the Unitarian pulpit we have had a series of discourses teaching the people your theory. Nothing brings out a crowd on Sunday, like the announcement that Darwinism is to be the theme. Surely the world moves!

Command me in whatever way you may wish observations made, on birds, insects, or plants, and I shall only be too glad to render assistance as far as in my power.

Accept my thanks for your courteous reply to my former letter, and believe me | Yours most sincerely, | Mary Treat.

> *Down,* | *Beckenham, Kent.*
> Jan 1. 73

Dear Madam,

I am very much obliged for your kind letter; & should esteem it a great favour if during warm weather next summer you will observe two points for me in Drosera Filiformis. Namely to place some flies within quarter of an inch of the apex of the leaf & observe whether it bends at all after an interval of a day or two. Secondly to rub with a clean needle a few of the glands with some little force, and to touch each gland half a dozen times; & then observe whether in the

Mary Treat, by Beatrice Braidwood, from a photograph
Courtesy of the Vineland Historical and Antiquarian Society

course of an hour or two the hairs or filaments bearing these glands become incurved. I am glad to hear that D. filiformis catches only small insects, as I suspected this. I have observed with care several other species of Drosera. Does the Dionæa [*Venus fly trap*] grow in your neighbourhood? If so I much wish to learn what sort of insects it commonly catches, more especially whether large or small kinds. I have sometimes suspected that its structure & movements favour the escape of small insects.

D.̇ Gray has given a rather free translation of what I said to him about nerves; and this related only to Drosera. I have found that by pricking a particular point in the leaf I can paralyse half of it; but I must make many more trials next summer before coming to any final conclusion.

With my best thanks | I remain Dear Madam | Yours very faithfully | Charles Darwin

P.S. I subscribe to the American Naturalist, so I am glad to say that I shall see your article—

Vineland, New Jersey.
July 28, 1873.

Mr. Darwin—

Dear Sir,

I write to inform you of the result of my experiments with *Drosera filiformis*. But first allow me to tell you of the immense quantity I found, in its fullest vigor, just coming into bloom.— On the morning of July 7th, I started for the country, some thirty miles distant, and before noon of the same day, was in midst of acres of this beautiful plant, standing so thick as to exclude almost everything else. I had only found it in limited quantities before this, and never such strong, vigorous plants as these. So I wrote you that it caught only small insects, but now found I was mistaken. Great Asilus flies were held firm prisoners, and innumerable moths and butterflies, were alike held captive until they died; the bright pink flowers and glistening dew-like substance, luring them on to sure death. After the death of the larger insects they fall around the roots of the plants as if to fertilize them, but the smaller insects remain sticking to the leaves. The leaf itself does not coil around the insect, the sticky substance being sufficient to hold it, and the more the fly struggles the more it becomes entangled in the filaments, and the sooner dies.

I carefully removed strong plants away from atmospheric agitation, and found they would bend toward a struggling fly. I pinned living flies within a quarter of an inch of the leaf, in less than an hour the flies legs would become entagled in the filaments. I then tried them three quarters of an inch from the leaves. The leaves bent perceptibly away from the light toward the flies, but did not reach them at this distance. I tried bits of raw beef with the same result. I could see no effect produced upon the glands by rubbing them with a needle; perhaps I did not understand just how you wished it done.

But the most perfect, active fly-trap among these plants is *D. longifolia*. In less than three hours a vigorous healthy leaf will fold completely around a struggling fly, and bits of raw beef will become so enfolded in the leaf as to be completely hidden from view, while mineral substances—dry bits of chalk, magnesia and pebbles made no impression on the leaves. I wet the chalk in water, the filaments soon began to clasp it, but soon unfolded again, leaving it free on the blade of the leaf.

I also experimented with *D. rotundifolia*. You are aware that the filaments around the edge of this leaf are longer than those on *D. longifolia*. I placed raw beef on the blade of this leaf, in three hours the inner filaments are curved closely around it, and the longer filaments circling the edge of the leaf were slowly curving upward; in twelve hours all the filaments were clasping the beef, the glands like so many mouths, seeming to absorb nourishment from the meat. While on an equally vigorous leaf a dry bit of chalk made no impression. The filaments slowly curve around raw bits of apple, but not so firmly as around animal substances, and are nearly three times a long in making the movement.

Some days the plants worked much better than others, whether it was owing to the electrical condition of the atmosphere, or whether it was due to the amount of moisture therein contained, I had no means of ascertaining.

I sent Dr. Gray a box of the plants, but he begs me to go on with the experiments and publish results, as he is too much occupied to give them any attention.

Dionaea does not grow in this state, it has only been found in North Carolina, but I hope sometime to be able to visit its locality, and study it in its native home.

Yours very respectfully, | Mary Treat.

Aug. 12. 1873

Dear Madam

I am very much obliged to you for having so kindly sent me an account of Drosera filiformis.

Your statements will be very useful to me in my short account of this species. I am familiar with what you state about D. longifolia & rotundifolia.

I have just lately been working hard for the last 2 months on the latter species; & before long shall draw up an account of their digestive powers & action under various stimulants. If I were in your place I should be afraid to publish the statement about the Drosera bending towards flies or meat which they did not touch; unless I had tried the experiment many times, under the most rigorous precautions; for I am convinced that no botanist w^d believe the statement unless all the precautions taken were described in detail.

With my best thanks, I remain dear Madam | yours faithfully & obliged | Ch. Darwin

Whenever my little book is published I will do myself the pleasure to send you a copy; it will be on the genera Drosera, Dionæa & Drosophyllum with a republication of my paper on Climbing Plants.

Vineland, New Jersey,
June 8, 1874.

Dear Mr. Darwin,

Some time ago you asked me some questions with regard to *Dionaea*. I was not at that time prepared to answer, but since the latter part of April I have been giving the closest attention to these wonderful plants—now in their best working condition— I am with them during a large part of each day, while the insects are the most active. I have over thirty good, strong, vigorous plants; twenty-five of these I have numbered, and keep a record of the closing of each leaf, and the kind of insect it captures, and the number of days before it uncloses, with many other items. The remainder of the plants I am working with, with a view to see if there is any other point so connected with the bristles on the upper surface of the leaf-trap—the seemingly nervous centre—so that I can make any perceptible effect upon this centre.

One plant has caught two of the sprawling rose-chafers (*Macrodactylus subspinosus*). These beetles are quite strong, and one of the fellows escaped from two traps, but was finally captured by a vigorous leaf that closed over him so quickly, there was no space left for his head to get through.

About two weeks ago a leaf captured a homopterous insect (*Metapodius nasulus*), nearly as large as the squash-bug (*Coreus tristis*). When caught it emitted a disagreeable odor, peculiar to this class of insects; and to my surprise the leaf opened yesterday in good condition, and there was nothing left of the insect but the shell.

There is something about these plants that attract insects. Honey-bees wander over the earth close to the base of the plant, sipping the moisture, this is not because of lack of moisture otherwise, and the plants are in close proximity to a flower garden, and beside, the lawn is now covered with white clover blossoms. So I was induced to examine the roots. The bulbs certainly exude a sweetish mucilaginous substance which the bees are after, but how wary they are of the trap, not one has yet been caught.

You asked what kind of insect Dionaea commonly caught. It most commonly catches Dipterous flies, frequently much smaller than the house-fly. If a fly is large enough to move a bristle, so as to close the trap, I never saw it escape from a vigorous leaf—one that acts quickly.

I am also continuing my observations on the Droseras, have added one more to my list, *D. brevifolia.*

If there is any other point you wish observed, in any of these plants, I will gladly do what I can for you.

Dr. Wood of Wilmington—the only district in which *Dionaea* grows—is cooperating with me in these experiments.

Yours most truly | Mary Treat.

Down, | *Beckenham, Kent.*
June 22nd 74

My dear Madam

I am very much obliged for your extremely interesting letter, & I am glad to hear that you are studying Dionæa in so earnest a manner My observations on cultivated plants are now complete, and I shall publish them in six or nine months; though they will be of little value compared with those made on the plant in its own country. As you kindly offer me information, I should very much like to hear about one point. Dr Canby says that the same leaf will catch 2 or 3 insects successively. Now I find with cultivated plants that a leaf which has once caught a good sized insect, though it will open & remain so for a considerable time, has so little power of movement that it most rarely is able to catch a second insect or to close over any object. I should very much like to be able to say, what the truth is on this head.

I remain dear Madam with my best thanks | Yours very faithfully | Charles Darwin

Vineland, N. Jersey.
Dec. 2, 1874.

Dear Mr. Darwin.

I have been studying the bladder-bearing species of *Utricularia* off and on the last year, and am now fully satisfied that they are the most wonderful carnivorous plants that I have yet seen. The so-called little bladders seem to be receptacles for digesting animal food. Not only small animalcules are lured into these receptacles, but animals large enough to be distinctly seen with the naked eye; and by holding the little bladders up to the light the movements of the animals can be seen with the unassisted eye.

I have found the remains of several Cyclops in these bladders, and one living one apparently just incarcerated. But the largest animal, and one that seems most constant in these receptacles, is a snake-like larva with brush-like, telescopic feet— I cannot now recall its name—it is quite active when first imprisoned, thrusting out and drawing back its beautiful feet, but it is in such close quarters that this is about all it can do, coiled around as it is—the larger specimens with head and tail meeting.

Last evening I found one just incarcerated in a very transparent bladder, it being the sole occupant. It was very active with its telescopic feet and horns, but in the morning—some twelve hours having elapsed since my last observation—it had no longer the power of thrusting its feet in and out, it could only move the brush-like appendages, and a slow movement was visible in the dark intestine that traverses the length of the body; and now this evening, twenty-four hours having passed, no movement is visible in any part of the animal, but it is slowly disintegrating.

This is the history of many specimens that I have watched in the same way.

I never knew, not even a small animalcule to escape after once inside the bladder.

I have not heard of any one making observations on these plants, and so I thought mine of sufficient importance to announce to you.—

Yours respectfully | M. Treat.

Darwin cited Treat frequently in *Insectivorous plants*, published in 1875. He was much puzzled by the action of *Utricularia* (bladderwort), speculating that small animals for some reason forced their way into the bladders. Treat suggested that the bladders opened when brushed by a small organism, and that the partial vacuum thereby created sucked them in.

Green Cove Spring, Florida,
Apr. 3, 1876.

Dear Mr. Darwin—

I came to Florida in November last, and have been working on the carnivorous plants here. With this letter I send you pressed specimens of the Pinguiculas which I have worked with. I shall soon publish my observations, and will send them to you in print. I sent you Harper's Magazine for February containing my article entitled "Is the Valve of the Utricularia sensitive"?

I think I have found two distinct species of Utricularia since I came here, one growing in a warm sulphur spring in beautifully clear water, this species has no antennae, but it is not at all like our *U. purpurea*, which you have noticed in the article I sent you.

But my *greatest find* has been a new water lily. It is really astonishing how it could have escaped the botanists. What have they been doing to let me come down here and find this beautiful lily? I enclose specimen of leaf. There are acres of it in extent growing in the bays and coves of the St. John's river. It is one of the most beautiful plants I ever beheld, and when I first saw it my heart fairly stood still. It cannot be a *variety* of *Nymphaea*, but a *distinct species*. The character of the plant is unlike our Nymphæa, and it produces large double yellow flowers. I have sent the plant to Dr. Gray, and asked him to give me directions to send it to Dr. Hooker. I sent it to

Dr. Gray a week ago, but it takes so long for letters and packages to go from one end of the union to the other, that I grow impatient and write to you before hearing from him. If you are in communication with Dr. Hooker, please tell him about this water lily. I have transplanted it, and know that it will stand pretty rough treatment. It sends out runners, and even the little plants on the ends of the runners grow readily.

I have just met one of your countrymen and his charming wife— Mr. and Mrs. White who are traveling in this country. Mr. White is a member of Parliament and has traveled with Dr. Hooker. Mrs. W. is a good botanist, and is drawing and painting our Flora. I accompanied her in a row boat to this bed of water lilies, and she is to paint it for me.

Dr. Gray thinks that I have also found a distinct species of *Amaryllis*, it blooms some two months earlier than our *A. atamasco*, and the leaves are much longer and broader. It commenced blooming early in January and the leaves and flower-scapes are now dying down, and the bulbs are ripe, whereas our *A. atamasco* is now in full flower.

I remain here until about the 10th of May, when I shall return to to Vineland, N. Jersey.

Most sincerely yours, | Mary Treat,

P.S. As soon as I can press good specimens of the entire plant of the water lily—flowers roots & runners, I will send to you & Dr. Hooker if you desire.

P.S. You will not get a very good idea of the beauty of the leaf of the water lily from the pressed specimens, when fresh, it is very glossy, and finely blotched with red.

I enclose some of the larger leaves in the package with the Pinguiculas.

M.T.

Down, | Beckenham, Kent. | Railway Station | Orpington. S.E.R.
April 21st/76/

Dear Mrs Treat

I congratulate you on your splendid Botanical success in finding the Water Lily & Amaryllis. Alas I know nothing of systematical botany. The specimens which you have sent of Utricularia are most beautiful & excellently preserved. I shall feel great interest in reading your account of them when published.

I am sorry to say that I never received the article in Harpers on the sensitivity of the valve in Utricularia,— a subject which drives me half mad.— If you have been able to prove either side of the case, I beg you tell me exact title & date of the number, which I can then easily procure.

If the Nymphaea is presumed by Dr Gray to be a new species, I am sure Dr Hooker w^d be very glad of a specimen.— I hope that you received my Book on Insectivorous Plants, copies of which were despatched at the same time to you, Asa Gray & Dr Canby.

Dear Madam | Yours very faithfully | Ch. Darwin

Green Cove Spring, Fla.
May 15, 1876.

Dear Mr. Darwin—

Your most welcome letter was forwarded me from Vineland. I am staying here longer than I had anticipated, in order to make good and abundant specimens of the water lily to which I refered in my last letter to you. It did not commence blooming until about the first of May. It seems that Audubon found and figured this yellow lily in his book of birds of the South, but Prof Sargent—Director of the Garden at Harvard—tells me that "botanists had generally considered it a poetic fancy of the author, rather than a true delineation of nature", so it was not mentioned in any of our books of Botany. You may imagine how delighted they are at Harvard to get this lily. I shipped them a box of roots some time ago, and now word has come for 500 more roots. Dr. Gray has already sent a plant by mail to Dr. Hooker, and I will be pleased to send him pressed specimens if he would like.

Yes, I received your book on "*Insectivorous Plants*", and thought I had acknowledged its receipt. I was so fascinated with it that I sat up nearly all night before I could lay it down.

I think I have proven that the valve of *Utricularia* is sensitive, I will leave you to judge when you read the article. I have written to the Editor of the Magazine to forward you a copy, and it will probably reach you shortly after this letter.

I saw a large number of mosquito larvae caught in the valve with their heads left sticking out, and I put a great many such specimens in alcohol intending to send them to you, but my departure for Florida was quite unexpected, and I left them behind. If you would like them I will send when I return to Vineland.— The drawing of the mosquito larva (and the way it was caught) was made by Prof. C. V. Riley, from specimens I sent him.

I have been at work on *Sarracenia variolaris*, for some time past; it seems to me to be the most wonderful of all our insectivorous plants. My account of this will also be published in Harper's Magazine.

Dr. Gray asked me to publish the Sarracenia article in the *American Naturalist*, and you may wonder at my selecting a literary Magazine rather than a scientific one, but I am wholly dependent upon my own exertions, and must go where they *pay best*.

I start for the north tomorrow, it is very warm here—like July and August at the north.

Most sincerely yours, | Mary Treat.

Down, | *Beckenham, Kent.* | *Railway Station* | *Orpington. S.E.R.*
June 1st 76

My dear Madam

I have received your kind letter & the article which I have read with the greatest interest. It certainly appears from your excellent observations that the valve was sensitive, & I hope it may be so from homologising with Pinguicula; but I cannot understand why I could never with all my pains excite any movement. It is pretty clear I am quite wrong about the head acting like a wedge.— The indraught of the living larvæ is astonishing.— I do not at all understand the German account of the development of the utricles & I suspect that these Germans did not look to young enough utricles.

I am not well & am staying away from my home for rest, so pray excuse brevity. Wishing you success of every kind in your admirable work

I remain | Dear Madam | Yours very faithfully | Ch. Darwin

Sophie Bledsoe Herrick married an Episcopalian clergyman in New York in 1860; she had three children with him before they separated in 1868, after which she and the children lived with her father in Baltimore. She was head of a Baltimore girls' school from 1868 to 1872. She studied biology at Johns Hopkins University in 1876 and became fascinated with the microscope. She wrote articles on cell biology for the *Southern Review*, edited by her father, and succeeded him as editor after his death in 1877. In 1879, she returned to New York and joined the editorial staff at *Scribner's Monthly*, later the *Century*. A colleague described her as 'an intellectual woman of keen literary perceptions and of a scientific training and cast of mind'. She attributed Darwin's popular success to his lucid and approachable writing style. She wrote two science books for children (one on botany, one on geology), the introduction to *Essays and reviews of George Eliot*, and *The wonders of plant life under the microscope,* in which she discussed Darwin's botany, in particular his work on insectivorous plants, and mentioned Mary Treat's work on *Utricularia*. She was deeply religious, believed in duty, and respected Darwin as a scientist.

34 McCulloh St. Baltimore Md USA,
Feb 12th. 1876.

Charles Darwin, M.A. &c.

Dear Sir,

I have read with the most intense interest your volume, lately published, on *Insectivorous Plants*. The positive testimony is absolutely conclusive. But I want to ask you whether you have ever experimented upon these plants in order to determine whether they can sustain life, as other vegetation does, only upon inorganic matter. Whether, in fact, this wonderful power of assimilation is only supplementary to the ordinary powers of vegetation, or takes in any degree its place. "It appears, therefore", you say on p 18. of the Appleton Edition, "that the roots serve only to imbibe water". Is this because there is only water to imbibe, or because they lack the normal power of roots?

In very carefully examining portions of Dionæa leaf through the Microscope it seemed to me that the sessile glands must be morphologically stomata; by throwing the object a little out of focus they look exactly like the stomata on the back of the leaf, both belonging to the epidermal systems, it seemed not improbable that this might be so. Have you ever studied the development of the glands? Mirbels observations [*in Mémoires de l'Académie des sciences de l'Institut de France (1835)*] upon the gemmæ of Marchantia polymorpha, that whichever side happens to lie uppermost developes stomata and whichever undermost root hairs, seems to give some color to the supposition.

Enclosed please find addressed envelope, if my questions are not too silly to deserve an answer, will you give me one in a few words; if they are, or you are too busy to spare the time pardon the intrusion, and the questions which are inspired by a real desire to know, not mere idle curiosity

Yours truly & respectfully. | Sophie B. Herrick.

Down, | Beckenham, Kent.
March 6. 76

Dear Madam

My chief reason for believing that Drosera drew the greater part of its nourishment from captured insects was its growing where no other plants could grow. I began experiments on the comparative growth of plants with and without insects, but failed by the plants being accidentally poisoned. It seems to me likely that the chief difference would be in the production of seeds. I did not observe the developement of the glands as this had been partly done by MM. Grönland & Trécul; but Mr Bennett has recently read a paper on these glands which is published in the Monthly Microscop. Journal Jan 1876. I can hardly believe that they are modified stomata.

I am glad my book has interest you.

Dear Madam | Yours faithfully | Ch. Darwin

6   Companion animals

Women were less active in the field of zoology they were in botany. Many of the observations Darwin was sent were from amateur observers, rather than from people who had made a special study of the subject. Most people in the Victorian period lived closely with animals; horses were used for transport, and dogs and cats might be pampered pets or semi-independent household companions; even an invalid might have a bird for entertainment. Darwin's comments on animal behaviour, in *Variation under domestication, Descent of man*, and *Expression of the emotions*, attracted much interest. The analogies Darwin drew between human and animal behaviour struck a chord with many; and women in particular drew conclusions about human treatment of animals. Where Darwin's female correspondents made formal studies of animals, they were on insects, barnacles, and earthworms. The correspondence on animals has therefore being divided into two chapters: Companion animals and Insects and angels, 'insect' being the popular term for any small, apparently insignificant, creature. (No reflection is made on the actual companionableness of insects—John Lubbock had a pet wasp— or on the usefulness of companion animals; the distinction, which is admittedly loose, points only to a difference of approach.)

The following account of learned behaviour in a dog came from Jane Loring Gray, the wife of Darwin's Harvard correspondent, Asa Gray:

> Botanic Garden, Cambridge
> Feb. 14— '70—
>
> My dear Mr. Darwin,
>   Dr. Gray says, "You write & tell about the dog!"— And indeed it was only a supposition of mine that he was suckled by a cat, from his queer tricks when he came to us, a young dog of about 7 months old— He then would chase his own tail for sport; but I have heard of other dogs doing that— But he still keeps up the trick of washing his face with his paws, & will sit as demurely as any old tabby, licking one paw & rubbing his face, & then changing to the other— I am glad if he has any tricks worth noticing, for he is a stupid little doggie at learning anything new, & has nothing but an affectionate heart & some beauty to recommend him— He came to us through one or two transfers, so I cannot know much about his puppyhood; but the lady who gave him to us is coming here this week, & I will ask her if she can find out anything about it—

…
Very faithfully | Yours, Jane L. Gray

Darwin's nieces were drafted for specific investigations into animal expression for *Expression of the emotions*. This letter to Lucy Wedgwood has to do with Darwin's research for *Expression*, which was published in 1872:

Down
June 8th

My dear Lucy

I hear that your Dog is a barker: please observe for me whether the (upper) lips are at all retracted or everted when he barks & just before he barks. My impression is that if you open a dog's mouth lips almost hide teeth; but that they are much more exposed, when he barks, which implies some contraction or eversion.

You are so good an observer that I know I can trust your conclusion. The Bark ought not to be a savage one, as that w^d give tendency to snarl— a joyful bark or bark of good spirits w^d be best.—

Think of any fact about expression of any emotion in any of your birds.—

Yours affect. | C. Darwin

Our Polly will not bark except as she rushes away to some supposed enemy in forest.

Anne Jane Cupples, who wrote children's books and was married to one of Darwin's most assiduous correspondents, George Cupples, a deer-hound enthusiast, wrote to Emma Darwin after reading *Expression of the emotions* in 1872:

The Cottage | Guard Bridge | Fifeshire
8th Nov

My dear Mrs Darwin

Please thank Mr Darwin for sending me his book, it was very kind indeed of him. I have not been able to do more than look at the plates, because Mr Cupples walked it off, and a minister here has got the promise of it after him, all without my leave in the matter. I have written to Aberfoyle to make inquiries about the dog who howls to the music   Miss Glen is a person to be relied upon, and I have given her instructions, and told her to be *very* particular. I cannot say for certain about the puckering up of the eyes or face   my *impression* is it simply winked its eye lids very fast, as one would do if hammering was going on, and that the eyes ran with or, were filled with moisture like a person crying, but Miss Glen will be able to tell us. After your letter came I remembered a young lady having told me their dog howled when particular tunes were played so I went off to see it, but after walking three miles I was told it was a dog they *once* had and that it was in Campbelton. She told me their ministers wife said their dog howled so off we set to get the dog, for in this case the minister had no piano, but we were told she ment her dog at home, her mothers dog in fact at Stirling. Then the minister who is a great wag began to tell me about a wonderful dog in St Andrews who used to pull the bell, and not only howl to music but played the piano and howled to his own playing. So I set off another day to find this wonderful animal. I found all he said about the dog was true, but he had been the property of Proff Macgill and the gentleman is dead and the dog is away   And now the Minister keeps teazing me about my novel hunt never failing to ask every time I meet him "Well have you caught the howling dog yet"   He has told several of the gentlemen in the neighbourhood too, who say "Oh by the way I have heard of a wonderful dog who howls, but he's in Sky." or some eaqually far away place. However though it has been a great amusement to *them*, it has been very useful to me for the stories I have heard about dogs, is something extraordinary throwing *my* poor dog quite into the shade, and I shall be able to tell my friends in Edinburgh who doubt my story. …

With my kind regards to Mr Darwin and many thanks once more I remain | Yours *very* truly | A J Cupples.

The author of the following letter was almost certainly Caroline Jemima Shuttleworth, the daughter of the warden of New College, Oxford, Philip Nicholas Shuttleworth. Her sister Frances Bevan is mentioned in the *Dictionary of National Biography* (*ODNB*) for her religious writing and translations of German hymns. When this letter was written (the year is unknown), Caroline was living with her widowed mother in a Hertfordshire village.

Wykeham Rise | Totteridge | N
Nov. 27

Sir

I am taking the liberty of writing to tell you of a curious instance of what appears to me like aberration of instinct (insanity?) on the part of a fantail pigeon, wh. on no theory can I account for. It is several years ago, before I was acquainted with y.r writings. I only wish that I still possessed the eccentric bird that I m.t add to my audacity in asking you to do it & me the honor of paying us a visit. But alas! it is no more. Still I have so often wondered how you w.d have accounted for its conduct, that at last I am constrained to write & ask you:

At that time we kept a few white fantail pigeons in a pigeon-house at the top of the coach house. One day I picked up somewhere an empty ginger-beer bottle—of the ordinary brown stone description, & I threw it, I dont know why, into the middle of the stable yard, just below the pigeon house.

Immediately the father of the fantail family flew down in a state of intense excitement, & to my great amusement began to perform the most extraordinary genuflexions, evidently in homage to the bottle.

He walked solemnly round & round it, cooing continually, & trailing his wing, & bobbing his head up & down, with the most exaggerated antics I ever beheld on the part of an enamoured pigeon. This went on for *hours*, but he never went quite up to it, & it never ceased until the bottle had been removed.

And this object never failed to attract him. Whenever an amusement was required for our visitors, I produced the bottle with invariably the same results. He flew down with quite as great alacrity, & usually far greater, than when his peas were thrown out for his dinner. The other members of his family regarded his performances with contemptuous indifference, taking no interest whatever themselves in any ginger beer bottle. I often tried him with other things, but only the bottle ever attracted him.

Now what was the cause of this infatuation? He c^d. not possibly have thought it was a pigeon. If it was insanity it was a monomania, for on all other points he was as sane a bird as you c^d. find.

With many apologies for troubling you with this anecdote I am
y^r. obedient servant | C. Shuttleworth

Darwin met Frances Power Cobbe at his brother's house in London. Cobbe was a journalist and a campaigner for women's rights and philanthropic causes in general and against vivisection. She lived with a companion, Mary Charlotte Lloyd. In 1872, Darwin wrote to Cobbe to praise her anonymous article in the *Quarterly Review* on the consciousness of dogs.

Down, | Beckenham, Kent.
Nov 28. 1872

My dear Miss Cobbe

I have been greatly interested by your article in the Quarterly. It seems to me the best analysis of the mind of an animal which I have ever read, & I agree with you on most points. I have been particularly glad to read what you say about the reasoning power of dogs & about that rather vague matter, their self-consciousness. I dare say however that you w^d prefer criticisms to admiration. I regret that you quote Jesse so often: I made enquiries about one case (which quite broke down) from a man who certainly ought to have known Mr Jesse well, & I was cautioned that he had not written in a scientific spirit. I regret also that you quote old writers; it may be very illiberal but their statements go for nothing with me, & I suspect with many others. It passes my powers of belief that dogs ever commit suicide; assuming the statements to be true, I sh^d think it more probable that they were distraught & did not know what they were doing; nor am I able to credit about fetishes.

One of the most interesting subjects in yr article seems to me to be about the moral sense. Since publishing the Descent of Man I have got to believe rather more than I did in dog's having what may be called a conscience. When an honourable dog has committed an undiscovered offence he certainly seems ashamed (& this is the term naturally & often used) rather than afraid to meet his master. My dog, the beloved & beautiful Polly, is at such times extremely affectionate towards me; & this leads me to mention a little anecdote. When I was a very little boy, I had committed some offence, so that my conscience troubled me; & when I met my father, I lavished so much affection on him, that he at once asked me what I had done, & told me to confess. I was so utterly confounded at his suspecting any thing, that I remember the scene clearly to the present day; & it seems to me that Polly's frame of mind on such occasions is much the same as was mine, for I was not then at all afraid of my father.

This note is not worth sending, but I have nothing better to write & I remain with kind regards to Miss Lloyd | yours very sincerely | Charles Darwin

Cobbe replied:

26. Hereford Sq. | S W

My dear Mr Darwin

You will know how pleased & proud your letter has made me.— I do not write to draw from you any more words, valuable though they be either as praise or as criticism, but just to say I have sent you the Cornhill [*Magazine*] just published to shew you another instance of what seems to me genuine superstition in dogs. Of course I rather turned the matter to jest in calling a stump a fetish—but the sentiment of *vague awe* at *the incomprehensible* is surely, I think, to be traced both in such freaks of dogs & in the shying of horses; & is very nearly akin to human superstition if not quite the same feeling.

Your own & your dog's similar behaviour after the commission of guilt, form delicious counterparts! Do you think the phenomenon can at all explain the exceeding *religiosity* of a great many arrant moral offenders?

Miss Lloyd entering agrees with you about the suicide of dogs & wished me not to insert the stories in my article— I did so in truth only hypothetically. Is there not some sort of radiate creature which casts off its own limbs, and strictly speaking causes its own dissolution when captured?—

I am sorry to hear of Jesse's untrustworthiness. No doubt old stories of natural history are little to be relied on for scientific purposes. They only shew what men then thought their beasts might do—

Would that those cruel old Jews had one such anecdote as that of Ulysses' dog in all their literature   It would have stopped a thousand Christian atrocities

[*Cobbe later wrote:* 'Had it but been recorded of any eminent canonical Prophet or Apostle, as of the virtuous (but alas! apocryphal) Tobit, that he had had a Dog which followed him on his pious journeying, the fate of all the dogs in Christendom would have been improved' (Cobbe 1889, p. 39).]

We have just returned from the zoos where we paid a domiciliary visit to the Chimpanzee in his private study. The poor dear little beast took a fancy to me & stroked my dress & face affectionately. I must say the grasp of his strong warm hand,—gentle and cordial as any human handshaking, was quite awful to me. I should consider it every bit as much murder to kill him as an idiot— The nearer one feels in pity & sympathy to an ape or an idiot the more I think the vague sense presses on us that some positive *thing*—(a thing we may as well call a "soul" as anything else,—) is missing; & not that it is merely a rudimentary stage of development which we behold— This impression does not come from any theory—indeed it does not fit any theories at all—but it is one which comes to me with very vivid insistence

With many apologies for troubling you with this long letter & with my (very childish) article in the Cornhill, | believe me dear Mʳ Darwin | most truly yʳˢ | Frances P Cobbe
Nov. 28

Athénaïs Michelet, a French author, wrote to Darwin for advice on her own writing. She had been educated by her father until his death when she was fourteen, after which she took the only degree available to women, a teaching certificate. While working as a governess in Vienna, she began a correspondence with the French historian Jules Michelet, and later married him. She urged him to return to his early interest in natural history, and they collaborated on a number of books. The following letters from her have been translated from the French. Her book on cats, *Les chats*, was published posthumously.

Paris | rue d'assas | 76—
17 May 1872

Sir

As my husband's name is known to you, I hope, through his great historical books, and our admiration for your genius is frequently expressed in several of his works, I come as a humble disciple to request the support of your counsel for a book on which I am occupied at the moment.

It concerns *cats*, my guests and favourites since my earliest childhood.

Brought up in the country, in a sort of Noah's ark, I spent my solitary youth observing and noting the impressions I received from my habitual companions.

My book has no scientific pretensions, however. It is to be a brother volume to the *Bird*, the *Insect*, etc., in which we [*i.e. Athénaïs and her husband*] tried together to give society people a taste for natural history and to develop their desire to familiarise themselves with it from the masters of the science.

I have read the chapter on cats in your book on *variations* most attentively, and much regretted its brevity.

I can confirm, sir, that half-Angora crossbreeds do reproduce with one another and with common cats.— In connection with this fact, perhaps I can interest you by giving you some details.— Some years ago I received a pair of cats. The male, a magnificent black and white Angora, nevertheless seemed to show slight intermixture.

The female was a true gutter cat, meagre, *short-haired*, and slender.

In the spring, contemplating a family, the spouses discreetly withdrew to the bottom of my garden. But a *smoky black* cat came prowling about, and had the customary duels with the husband.—

Three kittens were born from that union. I kept them all. These were their markings:— Though black and white like his father, the eldest had short hair and a look of his mother.—

The second, a female kitten, had by contrast a coat entirely of her mother's colouring, but with her father's long silky fur, which later developed the warmest and most delicate glints. I called her: The *blonde*.

The third, was a *totally black* male, not the *jet* black of his father, but the *dull* black of the outsider.

The singularity was that his fur, more *woolly* than silky, had at the same time the length and volume of an Angora cat. I named him *Pluto*. I kept him for 5 years. He died of a gunshot.—

From his first year he and his sister were mated. But she, suffering from the excessively precocious union, had a bad litter. In her weakened state, she showed no maternal instinct. The kittens died almost at birth.— But after this sort of miscarriage, her health regained the upper hand; she became remarkably beautiful; yet preserved some indefinable moral grace in her attitudes and movements from those early trials. She might have been a person.

The year after, at the moment when she was about to give us kittens by Pluto, (the father and the elder brother had already met an accidental death,) she was caught in a snare, and strangled.—

This proves the fecundity of crossbreeds, and also proves that *intentional selection* is not so difficult as one thinks.

In Paris, where one keeps cats to which one is attached indoors, the couple, where there is a pair, seems to be completely uninterested in looking elsewhere. I have even confirmed that the spouses display signs of reciprocal attachment in the *calm seasons*. I have seen the female becoming irritated and boxing the ears of a husband whom she thought flighty.

The indomitable independence of cats and their pressing need to roam at night will always be an obstacle to the longevity of individuals in towns; but, by keeping the couple *sedentary* at the desired time, one can modify a breed up to a certain point.

Since your book was published before the two exhibitions held at London [*the first two official London cat shows in 1871*], you were not able to speak of them.

May I dare to enquire, sir, whether *reports* worth dwelling upon have been written? I should also like to know whether there are any biographies or partial studies by *amateurs* written in English on domestic cats and big cats, *lions*, tigers etc. Everything must reach you, sir, as if in tribute, and I am sure that you have this information which is so precious to me at hand.—

With our highest consideration, I remain yours faithfully | A Michelet

P S. I shall be honoured to send you my little book, requesting all your indulgence towards it in advance. That is a generous currency which great minds cannot withhold.

Darwin replied:

Down, | *Beckenham, Kent.*
May 23 1872

Madam

I am much obliged for the honour of your interesting letter; & it will give me much pleasure to do any thing that you wish, as far as lies in my power. I wrote immediately to a gentleman who had served as one of the judges at the Crystal Palace, for the dates of the reports, & for information about any books on cats.

I heard from him this morning that he was just starting on a journey, but that he w^d write to me in a week's time. As soon as I hear I will endeavour to get what you desire, & send you any publications which I can procure. I do not know of any works exclusively on the larger Feline animals; perhaps as good an account as any will be found in Brehm's Thierleben [*Animal life, a multi-volume illustrated German work*], which I believe has been translated into French.

I am much obliged for your information about the fertility of crossed Angora & common cats. It is not believd by physiologists that the characters of two fathers can be transmitted to one & the same individual; & experiments on plants seem to negative any such belief. But I was reading the other day a paper by Fritz Müller which seemed to render this belief in some degree probable.

I thank you for the kind present of your book, which I shall have great pleasure in reading when I receive it. Possibly I may find something about the expression & gestures of cats under different emotions; & this is a subject which much interests me.

I have the honour to remain with much respect | yours faithfully | Charles Darwin

Paris | rue d'assas 76
26 June 72—

Sir,

My excuse for having waited so long to thank you, lies entirely in my husband's state of health.— The heart disease from which he has suffered since our horrifying events [*the Franco-Prussian War*] has worsened. I am absorbed and broken, I live in terror of the future.—

Work under such conditions is nearly impossible. Life is too disrupted.— I am trying to regain my self-control during the brief respites that his illness allows me; but there is no freedom of mind.—

If I can get back to the subject, I shall profit from your counsel, and endeavour to state my impressions as they came to me when I was not thinking at all about writing a book.

What almost always impairs our judgement is that we do not keep a sufficiently close eye on our instinctive tendencies.—

In the case of cats, for example, there is no middle ground; no-one can be indifferent to them. Cats are loved or hated—a consequence of temperament?— That provokes a priori affirmations.— From that point onwards, everything is compromised. I don't conceal from myself the sort of ardour I have for them. Being myself a nervous creature, I can perhaps sense more and better what predominates or rather dominates almost entirely in them—

In the order of affections, one might say that the dog loves with all its heart, *the cat with all its nerves.— An infinity of detail in feelings.—*

This fact, particular to the feline race, does not rule out individual uniqueness and character differences, just as in humanity.— thus, I am writing one chapter with the following title: "There are *cats and cats.*"

Thank you again Sir for your excellent and encouraging letter, as well as for the pamphlets.—

Now I shall make a request of you. I should like to have a photograph of the man who has gripped me so much with his writing. If you sign it, dear Sir, you will be doubling its value.

With admiration and profound esteem, | A Michelet

The following letter was probably written in response to *Expression of the emotions*: the author, Dora Roberts, is otherwise unknown. The letter was kept by Darwin, and Dora's description of a horse screaming was added to the second edition of *Expression*, edited by Francis Darwin and published after Darwin's death. The observation was attributed to 'a lady', which suggests that either she chose not to have her name mentioned, or that Francis was unable to contact her to find out what she preferred.

The Greenways | Leamington
Decr 17th.

Charles Darwin Esqr
Sir—

May I trespass on your time while I try to relate a curious instance of misdirected maternal instinct which occured in our Hen House here?

A cat came to the cook mewing piteously and expressing both grief & excitement— The woman allowed her to pull her gown & then followed her to the Hen House where a hen—which had been very indignant because not provided with eggs to sit upon for some time past—was found in possession of two small & starving kittens— These she defended by beak & wings until fairly beaten off with a stick— She flew with fury at the mother cat especially— Some hours after the supposed restoration of the kittens the mother again came to her friend the cook exhibiting even greater despair—& on yielding to her entreaties once more the cook found that the hen had managed to convey the kittens to the very highest shelf of the Hen House—when a ladder had to be fetched in order to release the kittens again   How they had been carried to such a height we never knew— The instinct which drew the cat to the cook was odd as the woman disliked cats & had never treated her kindly— It seemed her sense of justice to which the creature appealed   With regard to the almost human scream of a horse in agony having heard it once I can never forget it   In a crowd in London the horse fell & got under the wheel of a carriage the sounds rang in our ears for days after as the most expressive of agony we had ever heard— The cat & hen adventure occurred at Collin House six miles from Belfast—

Yours obed^{ly}— | Dora Roberts.

A friend of ours was pursued by a pig once with wide open mouth Her description made the animal seem very terrible— A niece of mine can distinctly move her ears & draw them forward she cannot explain how it is done—

Pauline Perfilieff (as she signed herself, anglicising or possibly gallicising her name: she is Praskov'ja Perfil'eva in a modern transliteration) was the wife of the vice-governor of Moscow, and a second cousin, once removed, of Leo Tolstoy. This letter is translated from French.

Moscow
Febr. 22/6. 1874.

Sir!

Having had the pleasure of reading several of your works and having just finished (in the Russian translation, I could not get it in

English) the one on "the expression of feelings", which more than charmed me, I have made up my mind to importune you with a request which may seem a little eccentric to you, but you depict the "involuntary movements" so well and so comprehensibly for us amateurs who are not men of Science that you will understand me and be indulgent, to the point of sending me your carte de visite, which I ardently long to possess.

As I love animals with a passion, especially bulldogs, whose character, habits and attachment to their master (an attachment which does not vanish even in a rabid state, which I have unhappily had the opportunity to observe very closely, since I have lost three of my favourite dogs to it at different times) I have studied thoroughly, I am, begging your pardon, astonished that among your remarks I have not found anything on that interesting breed, which to be sure merited all your attention, given the mobility and variety of their expressions. Here in Russia, they have been made into companion dogs and have not been disappointing, thanks to their intelligence, temperament, mischievousness and fidelity. There is no animal more amusing.— The bulldog, that clown, a monkey by vocation, is the only one who can compete with comedy.

Requesting your amiable indulgence yet again, and without giving up hope of receiving a brief word in reply, I beg you will believe me, Sir, to be one of your most assiduous followers.

Pauline Perfilieff | née C^ess Tolstoy.

Thereza Mary Story-Maskelyne was the wife of Nevil Story-Maskelyne, professor of mineralogy at Oxford University. She had been born into a scientific family in Wales, and studied photography and astronomy at home as a young woman: her father, John Dillwyn Llewelyn, a pioneer in photography, built her an observatory. She also had an collection of botanical specimens, and knew George Bentham, one of the foremost botanists of the time. He deposited her botanical logbook with the British Association for the Advancement of Science. She acquired copies of some of Darwin's books and papers through John Lubbock, a friend of her husband and of Darwin. Although an invalid at the time, she did some work with a microscope. In this letter she responds to a letter from Darwin in the magazine *Nature*, asking for information about birds biting flowers, apparently to eat the nectar. Darwin reported her observations in another letter to *Nature*, published 14 May 1874.

112, Gloucester Terrace, | *Hyde Park Gardens. W.*
May 4. 1874

Dear Sir

Your letter in "Nature" vol 9. p. 482, leads me to think you may like to know that we have a Canary which is in the habit of flying about the room, & is as fond of primroses as wild birds seem to be.

'Thereza Llewelyn and dickies'. By John Dillwyn Llewelyn.
© The British Library Board, Photo 1246/1 (11).

I cannot keep a pot of these flowers in the room, for the bird attacks the buds & open flowers directly it is let out, nipping them off without touching the leaves. I am sorry I made no observation as to the portion of the flower selected; but I have enclosed the remains of Cowslips which the bird is very fond of eating—& which if left long enough in the cage, would disappear entirely; the stem being a very quickly disposed of portion. In each of these heads of flowers the same portion has been taken first—a bite into the bottom of the tube of the corolla.

N°. 2. is the work of a Siskin in another cage; both the birds would have left nothing, if I had not taken out the specimens soon after giving them to them.

They have plenty of water & often green food—and often try the leaves in my flower pots—leaving the flowers alone; but it was a remarkable thing to see how the Canary would find out my Primroses never mind in what corner I hid them.

I remain yours faithfully | Thereza Story Maskelyne

Gould Anne Wolfe was the widow of the Irish clergyman Charles Wolfe of Kildare. She seems to have come from a fairly well-off family, and as a girl travelled in Europe with her mother and sister.

81 Uppr. Leeson St— | Dublin
March 9. 1875

Sir,

I have just been reading with the deepest interest y.ʳ "Descent of Man" It has been the means of rendering many nights of suffering more endurable

   …

Will you think me very audacious if I venture to say, I seem to see a greater likeness between our dear faithful companion the dog—when arrived at maturity—& ourselves, than even between the embryo wonderful as that is. I think dogs *are* capable of *deep thought* or contemplation— they seem at times quite in a state of abstraction—. I have seen an old terrier apparently so lost in some deep meditation as to be quite unconscious that she was called to take a walk. & I have had to rouse he⟨r⟩ f. her reverie, when the pleasure shewed convinced me her unusu⟨al⟩ disregard of the summons was not caused by disinclination to move. Nor was she asleep at the time, for I had seen her *instinctively* move when the servant had brushed passed her, a liberty she always resented more or less according to her humor

I believe she not only reflected but harboured many passions feelings supposed not to be shared with us by animals— On one occasion I threatened her with a light driving whip— I simply laid it across her back it c.ᵈ scarcely have tickled her, she quietly & deliberately walked out of the hall door & hid herself among the laurels, just to vex me —& it was not till 12 oc! at night that the groom found her.

Her attachment to me was extreme, but for ten days after this, she took no more notice of me than if I had not existed She took no food offered by me, she ignored me, & never *seemed* to see ⟨o⟩r hear me— She kept up this punishm! till one day, happening to meet her on the stairs, I sat down ⟨b⟩elow her put my arms round her & ⟨tried⟩ to coax her— she resisted at first, but when by degrees she felt herself drawn into her old resting place, her delight at finding herself there was too strong for her she suddenly relented & for many days did not let me out of her sight for a moment, but clung to me with greater fondness than ever—.

She seemed to act with the deliberate intention of punishing me— That same terrier having been beaten by a servant, for chasing the hens when a pup, on becoming a mother used to sit on a step overlooking the poultry yard, to watch her pups who early shewed her fondness for the chase—& when they transgressed she w! catch them by the hind leg & drag them away thus teaching them herself—in order to prevent their being beaten— Surely here reasoning power was displayed—not called forth by any accidental circumstance arising at the moment—but connected with what had taken place *long* before—& w! must have been remembered & *thought* over carefully & in her treatment of me there seems to have been not only anger, for what she looked upon as an indignity, but something in her— w! resented what she must have *argued* out to herself to have been an act of *injustice*. I *thought* she had been out all the afternoon—& had often seen her punished for a like offence— it turned out she had only been at the gate with the groom— While speaking of animals resenting supposed injuries I cannot but call to mind an instance of a favourite thoughbred mare— She was perfectly gentle, but fiery if not carefully handled. She was trained by our own groom a peculiarly gentle man with animals, who had never touched her with a whip, one day he happened to have one when exercising her & on her refusing to stand for a moment when near home, he gave her one cut with it. At the time she merely gave a slight brisk jump, but when in her loose stall— after the saddle & bridle had been removed, just as the groom was leaving the stable—she *deliberately walked after* him seemed to measure her distance turne⟨d⟩ & kicked him, without any show of temper— her ears were not put back. Nor did she repeat the kick nor c! she have intended to hurt him *much*—as if she had— she might ⟨h⟩ave injured him severely— This was the *only* time the mare kicked him or anybody else. Was not this the fruit of thought? I cannot but think too that dogs *have* some kind of ⟨c⟩onscience— by conscience, I understand not that, w! *teaches* us what is right, but that within us, w! tells us whether our actions are in accordance with those laws w! we believe—whether rightly or wrongly—to be binding on us.—

We had a water spaniel of rare intelligence, he understood Italian *well*, as well as English, & we had not had a french servant long till he understood him perfectly— this dog was an unmitigated thief & very self willed— Although he knew a beating w^d. surely follow a theft, he invariably came to tell us of his delinquencies & it was his confession w^h. generally first brought the loss of the meat or fish to our knowledge, he w^d. *crawl* into the room & put himself down at my feet, sighing audibly.

I fancy on these occasions the beat⟨ing⟩ was of a mild character, yet it seemed to hurt his feelings so mu⟨ch⟩ that he w^d. retire into private lif⟨e⟩ for the rest of the day & remain below stairs in company with the servant w^h. he *never* did any other time. He used to lay regular traps & devise strategems to get himself let out— his favourite plan was to *retch violently*. When all the occupants of the drawingroom w^d. fly to the door— once outside his whole demeanour suddenly changed—& as *plainly* as *dog* could do, he laughed at us all as he scampered off out of reach—generally stopping at the foot of the stairs to give one defiant bark. I do believe he felt genuine sorrow— when his thieving propensities got the better of him—& though on his return from his excursions we *never* beat him—having let him out ourselves—he used to appear horribly ashamed & if possible w^d. creep in unobserved— I c^d. fill vol^s. with the extraordinary traits I have observed in animals, those ⟨of⟩ y^r works I have read have interested ⟨me⟩ very much— Natural history in all its branches having always been a favourite study with me, during a long period of enforced illness— I trust this may plead my excuse for inflicting these lines upon you.

Would not yr influence do much to prevent the *wanton* & *unnecessary* torture of those animals that are our faithful friends & companions & that—*it may be*—possess higher faculties than most people give them credit for— *I* w^d. not purchase freedom f^m. my pain—& I know b⟨ut⟩ well—what pain is—at the cost of harming one poor animal whe⟨n⟩ subject to vivisection—unless under conditions, w^h. would render the experiment painless

I remain | y^rs faithfully | Gould. A. Wolfe

Since writing this, I have read with *delight one* passage in M^r. Gregs' 'Enigmas of Life'. He speaks of a *possible* future for *some* animals a belief w^h. as you know was at any rate not condemned by Luther

7 Insects and angels

Women also studied animals that didn't come into the category of companion animals: the correspondence chosen here has to do with barnacles, insects, and worms. (The term 'insect' was formerly loosely applied to any small or insignificant animal.) Although they might be collected for aesthetic reasons and studied purely for their own interest, the economic importance of these creatures should not be overlooked. Barnacles slowed down ships, insects destroyed crops, and worms were important to soil fertility. In this arena, female practitioners could be taken seriously: no one could risk doing otherwise, if it sounded as if they knew what they were talking about.

One of Darwin's earliest female correspondents on the subject of zoology (specifically, barnacles), and geology, was Mary Elizabeth Lyell, the wife of the geologist Charles Lyell and daughter of the geologist Leonard Horner. Leonard Horner had taken great care with the education of his daughters; when Mary was 5 years old he noted that she was getting on well in reading, and was learning geography. When she was 13, he was teaching her Italian, and when she was 18 he was congratulating her on her drawing, which she studied formally, and planning to study shells with her. Mary collected land snails on Madeira in 1854. She regularly travelled with her husband and presumably helped him with his work, although the extent of her contribution is unknown. The following letter to her from Darwin of 4 October 1847 suggests that she had her own collection of barnacles.

Monday Morning

My dear Mrs Lyell

I am much obliged for the Barnacles; the one marked Bergen is the right one; but it seems I must give it "locality unknown": I do not think anyone could have called it a Conia. You shall have your specimens back, but having now passed your new shell, I sh^d like to leave it, till I go over all the genera again, which will be sometime hence, but I will pledge myself that your shells are returned.

Thank Lyell for his note,—what an awful joke it would have been if we had all subscribed for a horrid calf's head! It will be grievous if the Coal Saurian turns out a fish; I will hope still that Agassiz's positive assertions may be disproved by bones, as well as footsteps.—

. . .

Mary Lyell. From K. M. Lyell, *The life, letters and journals of Sir Charles Lyell.*
By permission of the Syndics of Cambridge University Library

I have been a good deal interested in Miller [*Hugh Miller, First impressions of England and its people (1847)*], but I find it not quick reading & Emma has hardly begun it yet; I rather wish the scenic descriptions were shorter, & that there was a little less geologic eloquence.

With thanks, believe me, dear Mrs Lyell | Yours very sincerely | C. Darwin

The following letter was sent on 24 October 1849. Darwin evidently expected Mary Lyell to be competent in all Scandinavian languages.

Down. Farnborough. Kent.
Wednesday night.

Dear Lady Lyell.

I am going to beg a *very, very* great favour of you— it is to translate one Page (& the title) of either Danish or Swedish or some such language.— I know not to whom else to apply & I am quite *dreadfully* interested about the Barnacles therin described.— Does Lyell know Loven, or his address & title? for I must write to him; if Lyell knows him I w^d use his name as introduction; Loven I know by name as a first rate naturalist.

Accidentally I forgot to give you the "Footsteps" [*Hugh Miller's Footprints of the creator (1849)*] which I now return, having ordered a copy for myself.— …

I hope that you will not find the page troublesome & that you will forgive me asking you.

Pray believe me | Yours very sincerely | C. Darwin.

Mary Whitby was a landowner and naval officer's widow from Hampshire, where for a time she was a neighbour of Charles and Mary Lyell. Among her many interests, she worked on silkworms, hoping to encourage silk production in Britain and provide a source of income for the poor. Her experiments were successful enough to allow her to present Queen Victoria with twenty yards of crimson and gold satin damask, and in 1848, she published her *Manual for rearing silkworms in England*. The British Association for the Advancement of Science published two letters and one lengthier report from her in its reports of its 1844, 1846, and 1850 annual meetings at York, Southampton, and Birmingham respectively. At the Southampton meeting, she met Darwin, who was talking to breeders of domestic animals as part of his investigations into selection.

Down Farnborough Kent
Sept 2^d

Dear Madam

Your great kindness in giving me last year at Southampton information on the varieties on the silk-worm, makes me venture once again to trouble you. My question is a very simple one, and yet I am very curious to have it answered on the best authority.— Whenever I have observed the moths raised from silk-worm kept by children, the wings have been more or less crumpled, & I have been assured that they can never fly. Does this hold good, especially in France & Italy? If it does, can you, inform me, whether the males & females are equally helpless as regards flight? I presume that they are in the same condition, as our domestic ducks, & I should be extremely grateful for any information on this point.—

You were so kind last year as to give me hopes that you would try two experiments on hereditariness (a point on which I am particularly interested) in the caterpillar state: the first was whether the black eye-browed kind would produce black or dark-eyed caterpillar children: the second was to see if the very fat caterpillars (which I think you called Frales & which you described to me in a very laughable manner) would produce moths; & if so whether their offspring would be likewise fat & silkless.— I can really hardly say, how grateful I should be to know the results of such experiments; for in a work which I intend some few years hence to publish on variation, there will be hardly any facts in the insect world.

Will you permit me one other question, namely whether you have ever observed any difference in habits, such as in manner of crawling, eating, spinning &c in the caterpillars of the different breeds, which you have kept.— I am well aware I have much reason to apologise for thus presuming to trouble you, & I can only trust to your kindness to excuse.

Pray believe, dear Madam, with much respect. | Your sincerely obliged | Charles Darwin

Darwin wrote again in October 1847:

> Down Farnborough | Kent
> Oct 14th
>
> Dear Madam
> I am extremely much obliged by your kindness in sending me so capital a suite of male & female specimens. I am surprised at their not being a marked difference in the size of their wings: what you tell me of the wings of the female being smaller than those of the male, in their early growth is quite new to me. I hope to get a friend to enquire how these facts are with the species in India.
> I am sincerely grateful for your kind promise to make further observations, on the points which interest me next summer.
> Pray believe me, dear Madam | Yours truly obliged | C. Darwin

In 1849, Darwin thanked Whitby for sending the results of her experiments on the inheritance of markings:

Down Farnborough Kent
Aug. 12th

My dear Madam

I cannot express too strongly my thanks for the extraordinary trouble which you have taken in the interesting experiment, of which you send me the result.— I had given up all hopes of knowing whether peculiarities in the caterpillar state were hereditary, but now the point is amply proved: there is indeed a wide difference between a probability, however high & such an experiment as you have made.—

I am, also, much obliged for the information about the S. French caterpillar breeds; I was not aware the differences were so great.

If it would not be asking too great a favour, I sh[d] be greatly obliged if you would take the trouble to inform me, should you ever observe anything remarkable in the hereditary principle, or in the differences in structure or habits between breeds in the Silk-worm.— I dare not do more than hint my curiosity to know whether the Frales would prove hereditary,—ie whether it would be possible to make a breed with cocoons destitute of silk.— In the eyes of all silk-growers, this assuredly would appear the most useless of experiments ever tryed.—

Pray accept my most cordial thanks, & believe me with much respect, | Your's sincerely obliged | C. Darwin

Whitby's replies have not been found, but in *Variation under domestication* 1: 302–3, Darwin reported:

> Several years ago Mrs. Whitby took great pains in breeding silk-worms on a large scale, and she informed me that some of her caterpillars had dark eyebrows. This is probably the first step in reversion towards the tiger-like marks, and I was curious to know whether so trifling a character would be inherited; at my request she separated in 1848 twenty of these caterpillars, and having kept the moths separate, bred from them. Of the many caterpillars thus reared, 'every one without exception had eyebrows, some darker and more decidedly marked than the others, but *all* had eyebrows more or less plainly visible.'
>
> ...
>
> Cocoons are sometimes formed, as is well known, entirely destitute of silk, which yet produce moths; unfortunately Mrs. Whitby was prevented by an accident from ascertaining whether this character would prove hereditary.
>
> ...
>
> I was assured by Mrs. Whitby that the males of the moths bred by her used their wings more than the females, and could flutter downwards, though never upwards. She also states that, when the females first emerge from the cocoon, their wings are less expanded than those of the male.

Whitby died in 1850; a phrase from her church monument describes her as 'of masculine sense and feminine charm of person' (Colp 1972).

Margaretta Hare Morris was an American entomologist from Germantown, Pennsylvania. Although she had no formal secondary education, she joined the American Association for the Advancement of Science in 1850, and the Philadelphia Academy of Natural Sciences in 1859. She published on the Hessian fly and the seventeen-year locust; like Whitby, her motivation was partly economic, as these insects did great damage to wheat crops and fruit trees, respectively. Her published writing had a brisk, authoritative tone: this is from a communication on the Hessian fly (Morris 1841):

> Having completed a series of observations on an insect that has for years destroyed the wheat in the neighborhood of Philadelphia, I now beg leave to lay them before the Academy of Natural Sciences, with specimens of the insect in all its forms, from the egg to the perfect fly. To those familiar with Mr. Say's description, accompanied by Mr. Le Sueur's accurate drawings, given in the first volume of the Journal of the Academy, no doubt can arise as to the identity of the male insect now presented with the Cecidomyia destructor of Mr. Say; but the female differs materially in colour, her body being entirely black or blackish-brown; and the wings are destitute of the hairy fringe so conspicuous in the male.

Darwin evidently heard of her discovery of fish eggs adhering to the legs of a water beetle, and asked his friend Joseph Dalton Hooker to investigate (*Correspondence* vol. 5, letter to J. D. Hooker, 7 April [1855]). Darwin was interested since he thought the phenomenon might provide an explanation for how fish colonised bodies of water such as mountain lakes. This letter, found in the Darwin Archive–CUL and annotated by Darwin, was written by Morris to Hooker's friend, Richard Chandler Alexander.

Germantown
June 17th— 1855

Dear Sir

I received your welcome note of May 2[d] and with great pleasure give you all the information in my power relating to the curious history of the water beetle, (Dyticus marginalis) which conveys the eggs of fish from place to place on its fins— I remember distinctly, relating to you, during our pleasant walk, the incidents so interesting to me, and the solution of the dark question of how our Mountain Lakes were peopled by the Lake trout, when these lakes had no inlet above ground, and these fish are never seen in shallow water or in the outlets of these lakes.

My first impression that the D. marginalis was the agent that conveyed the eggs of fish from lake to lake, was received from a story told, I believe, by Kirby, of a M.ʳ Smith, in England who caught one of this species, in his study, and finding it to belong to the water, put it into a glass jar, where it lived for some time, on examining itt the next day he found eggs, floating in the water, which had been adhereing to the fins of the beetle, these eggs soon hatched, and became fish of the same species, which inhabited a neighbouring lake.

In 1846 I had the pleasure of studying the history of this insect, on the shores of several of our mountain lakes in north Pennsylvania, and found that it fed on fish and the fish roe that it found near the margin of these Lakes, destroying numbers of the lake trout which are found only in these inland seas, in *deep water*, and never in the outlets, most of these lakes are supplied with water by springs at the bottom, only, and have no communication with other lakes, empty-ing into rivers, where lake trout are never seen.

While on a visit in Montrose, Susquehanna Co Pennsylvania, I was presented with a Dyticus marginalis, which flew into a window, attracted by the light of a lamp, he must have flown at least three miles, as there is no lake nearer that town, but several of some miles in extent about that distance from Montrose— this specimen had no roe on it—but those feeding on the margin of the lake were covered with it—leaving no doubt in my mind as to the fact that they thus carried the eggs, from lake to lake and peopled them with fish that had no other means of being transported—

If this meagre account can throw any light on the subject be assured it will give me much pleasure, and will continue to persue the subject if it continues to interest naturalists, it will make me most happy to communicate with you on this subject, or any other ques-tion in natural history that may have fallen under my observation.

…

My Sister joins me in kind remembrance | and believe me | Very respectfully | M H Morris.
D.ʳ R C Alexander | London.

Darwin's annotations were somewhat ungracious: he wrote against the fifth paragraph, the one beginning 'If this meagre account …' in which Morris offers to pursue the subject, 'This is the most important statement in Letter', then emended that to 'only important statement'. However, the marine biologist and author of *Darwin's fishes* Daniel Pauly concludes that Morris's observations were probably correct and her hypothesis reasona-ble, adding: Margaretta 1: Charles 0. Pauly would have been backed up by Asa Gray in Harvard:

D.ʳ Alexander asks me to enquire of Miss Morris about a story of hers of seeing a beetle with eggs of fish attached. He says he has written to get information about it for you, but no reply comes. This Miss Morris is a good observer. I know her sister very well, & shall write to-day, & hope to get the statement of the facts for you, very soon.

Katherine Murray Lyell, the sister of Mary Elizabeth Lyell (the two sisters married two brothers), was primarily a botanist; in 1870, she compiled a geographical handbook of fern distribution. However, she evidently also had an entomological collection, and by 1856 was seeking advice on how to organise it and dispose of duplicates.

Down Bromley Kent
Jany 26th

My dear Mrs Lyell

I shall be very glad to be of any sort of use to you in regard to the Beetles. But first let me thank you for your kind note, & offer of specimens to my children: my Boys are all butterfly-hunters, & all young & ardent lepidopterists despise from the bottom of their souls coleopterists.—

The simplest plan for your end & for the good of entomology, I should think, would be to offer the collection to Dr J. E. Gray for British Museum, on condition that a perfect set was made out for you. If the collection was at all valuable I should think he would be very glad to have this done.— Whether any third set would be worth making out, would depend on value of collection: I do not suppose that you expect the insects to be named for that would be a most serious labour.— If you do not approve of this scheme, I sh^d think it very likely that Mr Waterhouse would think it worth his while to set a series for you, retaining duplicates for himself, but I say this only on a venture. You might trust Mr Waterhouse implicitly, which I fear, as rumour goes, is more than can be said for all entomologists.—

I presume, if you thought of either scheme, Sir Charles Lyell could easily see the gentlemen & arrange it; but if not, I could do so when next I come to town, which however will not be for 3 or 4 weeks.—

With respect to giving your children a taste for Natural History, I will venture one remark, viz that giving them specimens, in my opinion, would tend to destroy such taste. Youngsters must be themselves collectors to acquire a taste; & if I had a collection of English Lepidoptera, I would be systematically most miserly & not give my Boys half-a-dozen butterflies in the year. Your eldest Boy has the brow of an observer, if there be the least truth in phrenology.—

We are all better, but we have been of late, a poor household.—

Pray give my kind remembrances to Colonel Lyell & believe me, my dear Mrs Lyell | Yours truly obliged | Charles Darwin

Mary Elizabeth Barber was an artist, naturalist, and poet. Her family emigrated to South Africa with her when she was 2 years old, in 1820, and she was educated at home. Through letters, she became acquainted with Darwin and with Joseph Dalton Hooker, who as director of the Royal Botanic Gardens, Kew, was interested in information about South African plants. The only correspondence between Darwin and Barber that survives is Barber's responses to Darwin's questionnaire about human expression (see *Correspondence* vol. 15, letter from M. E. Barber, [after February 1867]). However, Darwin recommended her papers for publication. In this letter to the Linnean Society, he recommends publication of Barber's paper on the pollination of *Duvernoia adhatodoides*.

Report on Mrs Barber's paper.

Mrs Barber's paper seems to me worth publishing, because the dependence of the fertilization of a plant on one kind of insects alone, though not an unknown case, is very rare. Nor has any instance been recorded, as far as I can remember, of the access of other insects, being prevented by a mechanical obstacle requiring strength to be overcome. All the figures would of course be desirable and ornamental, but It would quite suffice to give on wood two of the adjoining flowers, one with a bee entering and one without. The outline fig 4 should also be given, as this shews better than fig 1 the contracted tube. It would be well to give the outline fig 3. if anyone can explain to the artist the structure of the fold embracing the base of the pistil, which is not clear to me.
Charles Darwin
To the Pres. & Council of Linnean Soc.
Down. May 10th. 1869.—

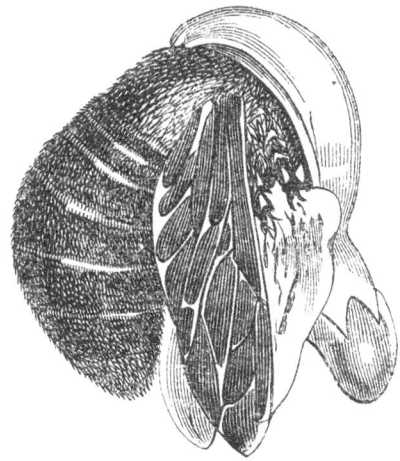

Bee entering a flower of *Duvernoia*, by Mary Barber.
From *Journal of the Linnean Society (Botany)* 11 (1871): 470.
By permission of the Syndics of Cambridge University Library.

In 1874, Barber sent her observations on the pupae of *Papilio nireus* (the green-banded swallowtail, a butterfly found in sub-Saharan Africa), which developed colours to match their surroundings. She questioned whether the colour change was brought about by the volition of the caterpillar, or whether the pupa automatically absorbed the colours around it, 'a sun picture or photograph'. Her manuscript was forwarded by J. D. Hooker to Darwin, who communicated it to the Entomological Society of London. Here is a sample of her writing from Barber 1874 (she had kept the cater-pillars in a case with a glass cover):

> When they had attained the full size of their species, and had ceased to feed, they at once set out upon their rambles in search of a suitable spot wherein to assume their dormant or pupa state; find-ing, however, that their travels were circumscribed, they appeared somewhat puzzled what to be at, and after a fruitless search for a 'leafy-dwelling,' several of them returned to the orange leaves, and there suspending themselves upon the small twigs, took up their com-mon form and colour [*dark green*]; others went to the bottle-brush branch, and there became pale yellowish-green pupae, of precisely the same colour as the half-dried leaves. One of the caterpillars in particular affixed itself upon the wooden framework of the case, where the wood and the brick came in contact with each other, and, to my surprise, this caterpillar, after throwing off its bright green skin, assumed the colours of *both the wood and the brick*, its under-side resem-bling that of the wood to which it was attached, and the upper side that of the adjacent brick-work.

Mary Treat, who appeared in chapter 5 as a botanist, initially approached Darwin about her work with insects. At the time she was still living with her husband in the progressive agricultural community of Vineland.

Vineland, New Jersey,
Dec. 20, 1871.

Mr. Darwin:
Dear Sir,

Experimenting with *Papilio asterias*, Cramer, I learned to distinguish the sex in the larva state—the female being larger than the male—and this led me to try to control the sex.

My first experiments were a a year ago last summer, some three or four hundred miles inland, where I had much better success than I had here last summer near the coast.

The larvae of my first experiment were of the first brood, so that I only had to wait a few days for their final transformation. These larvae fed on two quite dissimilar Umbelliferous plants—the Poison Hemlock (*Conium maculatum*), and the Caraway of the gardens (*Carum Carni*). I could always distinguish the larvae that fed on the Poison

Hemlock from those that fed on the Caraway; but with the butterflies there was little or no marked difference in their general appearance.

I noticed that the female larva fed longer than that of the male. So taking several larvae of the same age, I found some specimens were inclined to leave their food several days earlier than others, and these always proved to be males. It then occurred to me to try to induce some of these male larvae to feed longer; so, after they had wandered from their food, and even selected places for their transformations, (of course not *fixed*), if I disturbed them, made them leave their places, and coaxed them with a fresh supply of their favorite food, I could almost invariably induce them to eat from ten days to two weeks longer, when all such ones would be females— If it was a larva that fed on Caraway, I tempted it with the tenderest, and freshest leaves and flowers of the same plant. It was never hungry enough at that stage to be induced to change its food, though they will change in their earlier stages rather than starve, but evidently they do not like to change even then, and frequently fail to transform when their food is thus changed. They can be induced to change their food to the nearest allied species of plants with less difficulty—

On the other hand, when a larva had become the right size to produce a male, if I cut off its supply of food, even when it was eating greedily, it would wander about perhaps a little longer, as if in search of food, but finally it almost always changed to the chrysalis, and such a chrysalis always produced a male butterfly.

Last summer a fit of sickness prevented my experimenting with the first brood of these butterflies; and with the second brood—although I procured many specimens before the first moult—I was only able to rear about one in ten. An unnamed tiny Microgaster was their fatal enemy. But the few larvae that I succeeded with, behaved precisely as those of the previous year. Their food-plant was *Archangelica hirsuta*. I have the chrysalids of my last experiment carefully marked, and am fully confident that I know which will produce the different sexes.

...

I do not know that my experiments can be of use to you, but I thought perhaps they might interest you. A life time of observation and experiments could not repay the debt of gratitude we owe you.

Yours most respectfully, | Mrs. Mary Treat.

Darwin responded enthusiastically.

Down, | *Beckenham, Kent.*
Jan 5. 1872

Dear Madam

Your observations & experiments on the sexes of butterflies are by far the best, as far as known to me, which have ever been made. They seem to me so important, that I earnestly hope you will repeat them & record the exact numbers of the larvæ which you tempt to continue feeding & deprive of food, & record the sexes of the mature insect.

Assuredly you ought then to publish the result in some well-known scientific journal. I am glad to hear that your observations on Drosera will be published. …

I am very much obliged for yr courteous letter & remain dear Madam | yours faithfully | Charles Darwin

In December 1872, Treat extended her research to bees, challenging the belief that their sex was determined in the egg.

My observations and experiments with butterflies, lead me to think that the theory of the Hive bee is not correct. I know that I shall meet with opposition, so the only way is to experiment. I have already engaged a Langstroth observing hive for rearing queens, and shall carry on these observations, as well as continue my experiments with butterflies the coming season.

Darwin had asked the editor of the *Gardeners' Chronicle* to publish a short notice written by his niece Lucy Wedgwood on how worms pulled stones to cover their burrows (see *Correspondence* vol. 16, letter to M. T. Masters, 21 March [1868] and n. 2; the note, in *Gardeners' Chronicle*, 28 March 1868, p. 324, was signed 'I.W.', a misprint for 'L. W.'). Lucy continued her observations, which contributed to Darwin's final book, *The formation of vegetable mould, through the action of worms, with observations on their habits* (from now on cited as *Earthworms*). At his request, she marked out two square yards of ground at different locations, and measured the amount of earth that was thrown up by worms as worm castings over a year. In 1871, Darwin worked out the total.

Leith Hill Place. Dorking.
Nov 20th.

My dear Uncle Charles

Thank you very much indeed for sending me such a full account of the worm castings which I was deeply interested to read. What a wonderfully accurate guess of yours about the amount! The 16 tons an acre is the most astonishing part.

I have just been teazing the turkey cock— none of us have ever seen one *rattle* his feathers, wing or tail, but only scrape the wings along the ground, at the same time making a slight sound, apparently from the throat. What he did when angry with me was to puff out all his feathers, spreading tail and wings, but not in the same manner as when shewing off, indeed as Papa remarked quite the reverse, for instead of stiffening his wings, he let them hang down quite loose;— and there was *no rattling*.

I saw him make 3 attempts to pick up a grain of corn before he c^d. succeed, for his fleshy nose appendage!

Y^r. aff. niece | Lucy Wedgwood

In *Earthworms*, p. 165, Darwin cited her as 'a lady, on whose accuracy I can implicitly rely'. More worm-work came Lucy's way in 1872. Darwin was interested in the angle of worm holes, and also how the activity of worms degraded the traces of furrows in long-ploughed fields.

Jan 5

My dear Lucy

Supposing that you have leisure during next 2 or 3 weeks, will you have a try with straight blunt knitting needle to ascertain, whether on steep slopes the worms come to surface at nearly right angles to the slope, or at nearly right angles to the horizon.—

We have no steep grass-covered slopes here;— On nearly level surfaces the worms come up at all conceivable angles.— It w^d be *very important* for me if I c^d. ascertain that they generally come up at rt. $\angle$^s to the slope.— It is not easy to probe the holes.—

Yours affect | C. Darwin

The answer came on a postcard. Lucy was evidently also digging trenches to measure the disappearance of ridges and furrows on ploughland that had been converted to pasture (see *Earthworms*, p. 292).

6 Q. A. S^t.
Jan 20th

Out of 25 worm-holes probed with a blunt wire between Jan 6 & 14 on different slopes for a few inches, 8 came to the surface nearly vertical to the slope; the rest at various angles.

I will try some more, and when I have dug some more trenches will send the result about furrows

Down.—
Jan 21st

My dear Lucy

You are worth your weight in Gold.— I looked at a good many holes, but kept no account, & it tires my head stooping. It seems natural they sh^d come on average more often at right angles than oblique, to surface; but whether I shall be able to form a judgment I know not.— I shall be very glad to hear any further observation, & about furrows. It is at present all working in the dark.— I am now getting more inclined to trust the result of trenches cut across old furrows on nearly level surface; or to upper & lower part of grass-slope with no old furrows.

I have had some curious observations from Wroxeter, & William is working at Stonehenge for me.— I hope in time to come to some approximately safe conclusion.

If worms would be so good as to come up generally at right angles to slope, it would bring the earth down grandly. By the way I suppose when you say "vertical to the slope" you mean perpendicular or at right angles to the slope. The Mathematician George says vertical always relates to the horizon, so you ought to hide your diminished head.

Yours affectionately | C. Darwin

I find after the late heavy wind & rain the soft subsided castings are much 'blown over to leeward, even on level grass-field; the sections of all the recent castings were thus

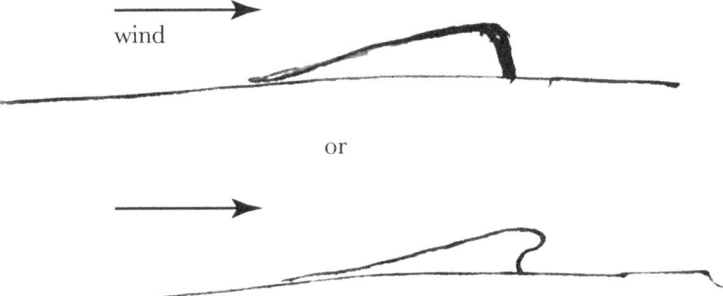

Would you visit the common on Leith Hill when you go home, & look at castings; the late storms must have blown there with terrific force.—

Lucy kept up her observations while abroad:

> Chalet de Villars | Montagnes d'Ollon | Aigle, Vaud
> Dear Uncle Charles.
>
> The steep grassy hill-sides here are very distinctly scored with ridges. They run parallel to each other with great regularity, about 3 or 4 ft apart and about 1 ft wide or nearly. They *very seldom* run into each other. There are worm-casts—but not abundant. I cannot help thinking they are made by the cows—(there are very few sheep)— they are undoubtedly used by them, being very often cross-ridged into those peculiar furrows cows always make in soft places by step-ping in each others footsteps. I thought it w^d be useless to draw them as it w^d be merely a n^o. of parallel lines.
>
> We stay here another 4^t night (till 29th) in case you (and the post takes 4 or 5 days) should wish to know anything more about them; If not, I need hardly say of course do not answer this.

We are enjoying ourselves exceedingly in this most beautiful place, with splendid weather

Y.ʳ aff.ᵗ niece | Lucy Wedgwood

After Lucy's marriage, Darwin tried to interest her sister Sophy in worms. The following exchange took place in 1880.

Oct 8th.

My dear Sophy.

Will you be so kind in any of your walks as to observe whether there any or many worm castings in the midst of Heath. It would be best to look where any grass-covered path crosses Heath, for if there are castings on the grass-covered paths or road & not amongst the Heath, it would show that heath is somehow unfavourable for worms. I ask, because I find a memorandum in my notes, that "there does not appear to be any worms amongst the Heath on Hayes common".— If Lucy is with you, I know that she would readily look from her well-known affection for worms— I am also becoming deeply attached to worms.— Can Lucy remember what sort of lantern she used when she looked at the worms. We find that the light frightens them.

Give my best love to your Mother; I do hope that she is better

Yours affectionately | C. Darwin

LHP.

Oct. 15.ᵗʰ

Dear Uncle Charles

I have been up on the common today after the worms.

I could not find anything in the middle of the heath, away from the paths, but there were some worm castings on the edge of the grass covered road, resti⟨ng⟩ on sprigs of heath, & in one case or more, with the sprays of heath pushed up through them. But there was grass or other plants all growing among the heath, in fact it is hardly pure any where.

I should have thought it a very unlikely place for worm casts, among a thick tangled mass, whether heath or not?

Also, from my experience, I shd hardly have expected to meet with any worms in that stony sand, (or in peat either.) unless in such a case as grass roots, weeds, in a gravel path. I dug a little among the heath, but found none, but this does not shew much, as I only in one case found any in digging the grass beneath actual wormcasts.

I don't know whether Lucy has written to you, she went up one day, & found nothing particular I believe, and meant to have gone again, but had to go home rather suddenly on account of one of the children not being well— She could not remember what light she took, unless she has been able to tell you since— I am sorry to have written you such a long winded statement about so little.

As my mother has been able to write herself, I will not about anything else. I do hope aunt Emma is better.

yr affect. niece KESW.

Mary Willes Tanner, wife of the London physician Thomas Hawkes Tanner, wrote to Darwin after reading *Earthworms*.

Providence Villa | Ventnor I. of Wight
12th. Dec^{br}. 1881

Dear Sir,

I have just read, with very great pleasure, your book on "Vegetable Mould & Earth-Worms". In connection with the large quantity of worms that live in a given space I think the following incident may interest you.

Late in the summer of 1872, towards dusk, after a very violent storm of rain, I noticed a peculiar appearance in my Garden, at Blackheath; and on going out I found the ground covered with earthworms. I may say without exaggeration that a net-work of these creatures covered the lawn, gravel paths & flower-beds, over an area of 80 feet by 30. It was impossible to walk without treading on a great many, and all were lying motionless, with the posterior ends in their burrows, with the exception of those that quickly retreated on perceiving the vibration of our footsteps. The garden sloped slightly *from* the house, and the worms were more numerous at the higher part, diminishing towards the lower level. They remained till it was too dark for us to see them: by the morning all had disappeared. The same thing was repeated after another storm, a few weeks later, only on a smaller scale; and never occurred again during the three years I remained at Blackheath.

I worked constantly in my garden, so that had there been evidence of a large quantity of worms I must have noticed it; on the contrary, I should say that the castings were under the average, and the number of buried leaves small.

Believe me, dear Sir, | faithfully y^{rs}. | Mary W. Tanner

Similar interest was shown by Ada Kepley in Illinois. In 1870, Kepley was the first American woman to graduate from law school. As a woman she was debarred from working as a lawyer until Illinois state law was changed in 1872; her husband, who had encouraged her studies, drafted a bill banning sex discrimination in professional occupations. She was a campaigner for women's suffrage and for temperance, and she ran for state attorney general of Illinois in 1881. She was ordained as a Unitarian minister in 1892.

Effingham Illinois. U.S.
Feb. 20th 1882,

Chas. Darwin,
Dear Sir

Hoping I do not intrude, I send you an item concerning "earth worms," bearing upon some discussions I now see concerning their alleged habit, of eating the roots of pot plants and injuring them by their acid excrements. I have had many house plants for years, and have kept my plants in unglazed, and glazed ware, in iron, in tin, and wood, have used garden soil, leaf mould chip dirt, and well rotted manure to enrich the same, and in which as you know earth worms abound   I have used no drainage by piling pot sherds, charcoal gravel &c, in bottom of pots, sometimes I have not even had a hole in the vessel.

I never bake the earth, always have worms in the pots, and never believed they were injurious to plants, have talked with women, who laid fallens to worms, but it could usually be traced to poor light, bad watering or high temperature, If I can have good light, proper temperature, good air, and water carefully, not too much nor too little, my plants will grow. I have grown all the ordinary house plants and some that are not ordinary, and with success, I think the worms do not hurt the plants.

I notice if there is some under the pots, they are fond of going out of the hole at the bottom, and making a sort of nest or bed

Very respectfully | Ada H. Kepley

8   Observing humans

In his research for *Expression of the emotions in man and animals*, Darwin approached friends, relatives, and correspondents for information. He had a questionnaire printed that his correspondents could send further afield, since he was interested in human expression around the world. Mothers in particular were asked for information about expression in infants. Often the responses to Darwin's questions are now as interesting for the light they shed on race, class, and the respondents' lives as for the subjects in which Darwin was principally interested.

Darwin's niece Frances Julia (Snow) Wedgwood contributed many observations, some of them drawn from her reading in Latin and Greek. The following, however, concerns the observations of her friend Jane Gourlay in a Lock Hospital. Little is known of Jane; she was tutor to the children of the nephew of Thomas Erskine, Snow's religious mentor, in Scotland, and taught at a school in Edinburgh. The Lock Hospitals were developed for the treatment of syphilis. (The name of these hospitals was derived from the Lock Lazar House of Southwark, but the origin is otherwise unclear.) They catered mostly for working-class women, and moral education would have formed part of the programme. It is not known which Lock Hospital Gourlay visited; if it was in a port or garrison town, the girl she visited might have been detained under the Contagious Diseases Acts, which allowed women suspected of being prostitutes to be detained and examined for signs of disease against their will. Darwin had evidently asked Snow for some literary quotations about the expression of shame; Snow sent a few via Henrietta, and then added:

> On the other hand (but I dont know whether it is to the point when Uncle C⟨h⟩ asks for quotations to give facts, & of course Uncle Ch cannot care for so very obvious a fact as that people *do* cover their faces for shame)— Miss Gourlay told me when I was asking her if her girls ever went wrong of one of them (they are all the lowest of the low) who had gone to the bad & ⟨w⟩hen she (Miss G) found her in a hospital ran away from her & hid her face & cd not be persuaded to look up till Miss G had to go away. Of course she was not one to reproach the wretched creature.

Darwin asked for more details, and Snow wrote again:

Friday night

My dear Harrot

Miss Gourlay answers me about that case of face hiding Uncle Ch wanted in detail, but there ⟨is⟩ almost nothing to say. She does not specially look after that class of women but had a little school for the lowest class of child⟨r⟩en & of course some of the⟨se⟩ fall into that way of life. This was a case of this kind. it was at the Lock Hospital she saw the girl, or rather did *not* see her, for she was in bed & as Miss G approached pulled the clothes over h⟨e⟩r face, & in spite of all her efforts to draw them off, would not allow her to do so,—so she cd not see any part of her face nor get any ⟨ans⟩wer from her. Afterwards on coming out she came to Miss G for help & counsel, but took offence because she gave her too thick & coarse a pair of ⟨bo⟩ots. That fact is irrelevant but I mention it to shew it was only shame made her hide her face.

Darwin described the case in *Expression*, p. 323: 'A lady informs me that she found in the Lock Hospital a girl whom she had formerly known, and who had become a wretched castaway, and the poor creature, when approached, hid her face under the bed-clothes, and could not be persuaded to uncover it.'

Another observation on a girl in confinement was recalled by Anne Barnard, the daughter of Darwin's Cambridge mentor, John Stevens Henslow. This time the subject was not expression, but animal-like ears. Anne would have been about 18 at the time of her observation.

Bartlow | Leckhampton | Cheltenham
March 30ᵗʰ 1871—

Dear Mʳ Darwin

Though I have not seen you since I was a child, the only excuse I shall make for troubling you with a letter, is that your "Descent of Man" has recalled to my mind something which I think may interest you. On Octʳ 20ᵗʰ 1852, I visited the Idiot Asylum at Colchester, with my father, Prof: Henslow— In the room of "Incurables", a girl was pointed out to us, who had very remarkable ears—"like a donkey's"—the attendant said, but that was an exaggeration. After the lapse of so many years, I cannot pretend to describe them very accurately, especially as the sights in the room were too horrid for us to care to remain there many minutes   The impression left on my mind is, that the ears rose to an unusual height, and were **pointed** somewhat like this

It is possible that there may be doctors or attendants still at the Asylum,

who may remember her, and be able to give you more accurate information about her, if you think it worth while to enquire, it is not *impossible* that she may be still alive.

I fear however, that there is—or was then, a limit to the number of years spent by any patient in the Asylum, and it is therefore very unlikely that you would find her there now—

Believe me to remain | Y^{rs}. truly | Anne Barnard

After reading *Expression*, which was published in November 1872, Laura May Forster (later to be E. M. Forster's aunt) sent this account of an alarming demonstration of rage to Henrietta: Henrietta forwarded it to Darwin.

Feb. 20th. 1873

...

I dont think I told you how strongly I was reminded of one of my Cannes boatmen by the bit in the expression book about shewing the teeth in anger— I cannot fancy a stronger instance of it; we were talking of the war & I was drawing the men out & rather baiting them with a careless sort of pity for the exiled Emp^r. [*Louis Napoleon (Napoleon III), former emperor of France*], a dark Italian looking man got increasingly fierce & scowling & said it was his own fault he was an exile—why did he not return, the traitor! to answer for his deeds like a man? I said ½ mockingly still, that was the last thing I blamed him for & I put it to the man what sort of a welcome he individually would give him? I declare tho' I heard no word of his answer it frightened me out of all jesting, he drew back & up the corners of his mouth till I saw every tooth in his head and to illustrate what he was say^g he clashed his teeth together so that my impression was he would fly at the Emp^r.'s throat like a bulldog. I was thoroughly frightened & yet so curious to know whether he really said what he acted, the whole boat was between us as he growled out the few words, that I asked him agⁿ. what he would do shew^g. I did not catch his meaning, he had been resting on his oars but then he pulled them in & using his hands to express his meaning with a savage grin "je lui torderais le cou comme ça" [*I would wring his neck like that*] giving a twist with his hands & shew^d. his teeth, but not clashing them as before. Somehow that seemed more human & I did not mind it so much, but the 1st. action with the teeth was appalling & I felt as if in danger from a wild beast. Even the milder edition expressed too much feeling for my taste & much as I like to study character I did not go out with my friends again. I felt as if I had played with edged tools & even at this time, after the lapse of a year to soften & sober impressions I cannot resist the conviction that I could chaff that man into drowning me, were he & I alone in a boat. I have often had to do with bad tempers & rough men but had no idea before how they might frighten me. I think that unconsciously both the man & I must have identified myself with the Emp! so that a portion of his wrath hit me as it were. …

Jane Loring Gray, the wife of Darwin's Harvard correspondent, the botanist Asa Gray, evidently had one of Darwin's questionnaires. While travelling in Egypt with her husband, after a lengthy visit to England during which the couple spent some time at Down, she sent these comments back to Darwin. The occasional numbers refer to the numbered questions on the questionnaire.

May 9— 1869—

My dear Mr. Darwin,

We are so grieved to hear of your accident! To one who has always to suffer so much, it seems hard to have any new cause added— We can only hope it will not aggravate old symptoms, & that we may soon hear that you are better—

I enclose the few notes I made for you on the Nile— I am afraid you will think us very stupid people not to have done more. But it is surprising when one's attention is drawn to it, how little we see *what makes* the different expressions in faces— We are satisfied with what we think the expression is— Then not understanding the language was greatly in our way, for the Arabs are great & excited talkers, & roll out consonants & their rs in a most emphatic way, & what seemed a quarrel or fierce discussion, turned out only an animated account of adventures— But they are a charming people! So docile & gentle, so courteous & dignified, such born gentlemen— Though our crew lo⟨ok⟩ed very odd to us at first as sailors, in their long, blue gowns & white turbans, one soon came to know each face, to admire such limbs as rarely greet European eyes— It seemed as if I had never seen walking & running before!— And the men keep up such perpetual ablutions as part of their religion, they never seemed like common sailors in coming in contact with them, & it was pleasant to have their willing help on all occasions. They are very much like children, pouting when vexed, & smirking like any conscious girl when they knew their beauty was admired; & some are so handsome!— Two of our crew were Ababdeh, a tribe of the desert towards the Red Sea— They were much darker than the Cairines & river natives, & quite different in feature, but the most gentle of them all, & such sweet smiles showing such White teeth!— As for blacks, the negroes from the interior were blacker than anything I ever saw before— A deep, lustreless, solid black.— But our crew burned & grew dark as whites do on our voyage— We thought when we first came back to white skins, how ugly they were! This lovely, golden bronze is so beautiful—

I thought you would have been interested in seeing an old picture here of Fra Angelico's of the deposition from the cross— The

Madonna has the distress muscles very carefully painted. The other women only the up & down wrinkles—

Pray give my love to Mrs. Darwin & your daughters. And heartily hoping you are relieved, | believe me very sincerely & cordially | Yours, | Jane L Gray

[*Enclosure*]

In Alexandria I saw the Arab commissionaire or dragoman of an old gentleman, who seemed a German, use exactly the gestures of *No. 13* [*i.e. the shrug*], when the gentleman would go in a direction different from that which he pointed to—

Passing slowly a common country cargo boat, the old man on board stood looking at us, with brow wrinkled & mouth compressed & upper lip raised— An expression, as I read it, of dislike & contempt— almost hatred—

A black towing a country trading-boat, when there came some obstruction in passing us, had the *distress* muscles, (the forehead transversely wrinkled in the middle,) very strongly marked, as he stood on the bank watching us—

I should say yes, certainly, to *No 6*— after watching a good deal— [*no. 6: When in good spirits do the eyes sparkle, with the skin a little wrinkled round and under them, and with the mouth a little drawn back at the corners?*]

(17) We all decided that the head was never "shaken laterally in negation" & that they did not understand what it meant— Nodding in affirmation was rare, it was more often a sign of approval or greeting.—

(2) One day the rais (Reis, i.e. Captain) of the small boat, in bringing up his boat nearly ran it into the large one, the men chaffed him a good deal, & my brother said he *blushed* quite to the back of his neck. He was a man full of gestures & very emphatic— When he wished to say he had nothing to do with a thing, that he washed his hands of it, he would hold both hands each side of his head, opened out flat—

One of the little village girls who carried our water-jar on excursions, a girl about 11 or 12, had one day some remark made to her by a young man of another party who passed us, evidently something which displeased her— She drew herself up, raised her head, compressed her lips dropping the corners, & fell back by one of the gentlemen of our party— Her expression was of great contempt— And the whole action very dignified—

(16) I heard the hiss to keep quiet, but as often from our dragoman & whites as from the natives—

A young woman sent Darwin her observations on her own blushing after reading *Expression*. The original letter is badly damaged; the text has been partly restored from the quotation of it in the second edition of *Expression*.

The Down Wood, | Blandford.
⟨   ⟩m⟨b⟩er 29ᵗʰ.

Dea⟨r⟩ Sir.

⟨Wh⟩en I am playing ⟨o⟩n the piano, and anyone ⟨co⟩mes, and looks over me; ⟨I⟩ am afraid they will ⟨loo⟩k at my hands, and ⟨I⟩ am so afraid of their ⟨being⟩ red that they blush, ⟨though they⟩—were not red [*before.*]

When my governess spoke of my hands ⟨   ⟩ being long or able ⟨to⟩ stretch, or drew a⟨ttention⟩ to them, they ⟨blushed.⟩ I once said this to ⟨   ⟩ I forget who, I think on⟨e⟩ of my governesses, she said "Oh yes, of co⟨urse⟩ everyone blushes in th⟨eir⟩ hands."

I hope you will not think I am trying to hoax ⟨you⟩ because I could ⟨not⟩ do such a thing, and ⟨w⟩ould be disgusted ⟨at⟩ anyone who did.

Yʳ truly | Maria Isabella Snow.

⟨I⟩ used to feel my hands ⟨ge⟩tting redder and redder, ⟨the⟩ more they were looked ⟨at⟩ I say they "used" ⟨   ⟩ise since I left the ⟨   ⟩ room, I never play ⟨   ⟩, so I have not seen it lately—

In *Expression*, pp. 314–15, Darwin remarked that blushing did not normally affect more than the face, neck, and upper part of the chest; he gave one instance of a woman's hands blushing. The first two sentences of this letter were quoted in *Expression* 2d ed., p. 333 n. 10, as being from 'a young lady'.

Darwin's interest in the age at which babies' eyes filled with unshed tears and the age at which they began to shed them led to a long series of letters from new mothers among his friends and relations. Pouting was also a subject he wanted to know about. Darwin may have been too tender-hearted to make babies cry himself: in a letter to Thomas Henry Huxley of 30 January 1868, he wrote: 'A dear young lady near here, plagued a very young child for my sake, till it cried, & saw the eyebrows for a second or two beautifully oblique, just before the torrent of tears began.' Darwin reported his findings in *Expression*, chapter 6, but did not identify the observers. It's evident from some of the letters below that Emma and Henrietta had been drafted to send letters of enquiry.

From Louisa Kempson, Emma's niece:

Plas Maur | Penmaenmawr | Conway
Thursday | 20 June 67.

My dear Aunt Emma

Will you please to tell Uncle Charles, that I have been making enquiries in my nursery about the tears. but I can only give him hearsay evidence as I cannot see so small a thing as a tear My nurse says that tears begin to stand in a baby's eyes when they are a few weeks old, & that they begin to run down the cheeks at about six weeks. my baby is just 4 months & the tears run down her cheeks in a piteous manner when she crys, which I am happy to say is very seldom of course I need not say that there never was such a baby since the world began! but I have never seen such a happy good tempered little soul—

From Cicely Hawkshaw, another niece, in 1868:

Hollycombe. | *Liphook.* | *Hants.*

Dear Aunt Emma

I am afraid it is too late to notice about the baby's tears with any accuracy for I have repeatedly seen her eyes full of tears already but can give no nearer date than that I must have seen them so before she was 3 weeks old; about the tears overflowing onto her cheeks I can observe as I have never seen it happen yet, indeed it hardly happens in what one may call babydom does it?

We are having such a nice holiday here and as all the tiresome shooting is over I have Clarke to myself and we ride and walk about and don't feel such strangers to the place as we did and the idle thoughtless life is doing Clarke good I am thankful to say.

Believe me dear Aunt Emma | Your affecte niece | Cicely M Hawkshaw

9th Feb.

From Elinor Bonham Carter in 1868:

March 20

Dear Mr Darwin

My sister in law Mrs Henry B C wishes me to tell you that the result of her observations on her youngest child up to this time is as follows: At 2 months & 2 days old—the nurse first saw its eyes full of tears, she herself saw the same thing at 11 weeks old— The child is now 12 weeks old— it does not weep tears, but the ⟨ ⟩ of screaming seems to force one or two tears out of the eye. If this is any use to you & if you can point out anything else to be observed, will you tell Henrietta as I shall see her tomorrow

I hope you are keeping pretty well & able to enjoy London. I was very sorry to hear that Mrs Darwin had the Influenza.

Believe me | yours affectionately | Elinor Bonham Carter

From Cicely Hawkshaw again:

Beverley
12 April 1868

Dear Aunt Emma

I observed tears running down Baby's cheeks for the first time on the 6th of this month when she was 3 months and 2 weeks old, the nurse says she had seen it before but I have been watching pretty closely and that was the first time I saw tears decidedly overflow; in the last few days too she has learnt to make her first voluntary movement and solemnly and with difficulty guides her hand to her own mouth.

Margaret Susan Vaughan Williams wrote after 14 October 1869. (The baby, Hervey, was to be the elder brother of the composer Ralph Vaughan Williams.)

Down Ampney | Cricklade

My dear Henrietta

I am afraid you will think I have taken a long time in answering your questions, but we do find it so hard to observe Baby's expression accurately, & besides I seldom *can* attend to his eyebrows, when he is crying and I am wanting to comfort him. I see my own eyebrow corners go up perfectly, and we are almost sure his do the same in a less degree—when he is going to cry—but the transverse wrinkles on the forehead not straight but waving downwards the lines converging towards the nose. I cannot draw a face in the least.

Baby crys with tears now and certainly has done so for a long while, but when he began I cannot tell.

He is just over 6 months old. I wish I could answer Uncle Charles questions better, but if we can observe further more accurately, I will write again—

…

With best love to Aunt Emma & Bessie | Your affec. | M. Vaughan Williams

Lucy Wedgwood, a regular research assistant, wrote to Darwin in January 1871:

Mary Owens 3½ yr. old child has a habit of sticking out her lips when she feels shy, but as it is *not* a pout of sulkiness I do not know if you care about it. She makes no sound. The lips do not seem to become tubular (that is the corners are not drawn together, or hardly.) The upper lip is stiffened and projected beyond the lower one, (tho' both stick out to a certain xtent) the lips sometimes not quite closed.

I have stupidly mislaid your paper about it & cannot find it. If I have forgotten any points I hope you will let me know, if you care about it.

Alice Mary Lane Fox, the wife of the archaeologist and anthropologist Augustus Henry Lane Fox (later Pitt-Rivers), wrote in 1875 to Ellen Lubbock, Darwin's neighbour, about her son's supernumerary thumb. It's likely that Ellen had mentioned the case to Darwin and he had asked her to make further enquiries. Darwin had discussed polydactylism in humans in *Variation under domestication* 2: 14–15 (published in 1868), but in the second edition (published in 1875) he withdrew his view that extra digits in humans were a symptom of reversion to a 'lowly organised progenitor provided with more than five digits'. Alice came across his comments on inherited extra digits that had the power of regrowth in *Descent*, published in 1871, and wrote this letter to Ellen in 1875.

Uplands | Guildford

July 25th.

Dear Lady Lubbock

   I trust I have been trained to take sufficient interest in all scientific investigations, not to think any question impertinent or unwarrantable the answer to which can in any way further that object. Indeed I felt very much inclined to write to Mr Darwin of my own accord when reading his interesting book on the Descent of Man when it first appeared on the subject. I am sorry however to say the Surgeon who performed the operation on my eldest son (Dr Trench Staff surgeon at Malta) has been dead several years but I will try as far as I can to make up for the more accurate & technical details he no doubt c^d have given. The extra digit or thumb was amputated by congelation in March /56 when my son was 4 months old. The excrescence was simply cartilage growing *a little above* the joint with a perfect nail as in the drawing I enclose which my son has just made of his hand as it was & is now with the regrowth which is entirely covered with nail quite loose from the bone— I don't know whether Mr Darwin heard that the inheritance was from my grandfather— J. T. late Ld Stanley of Alderley who died at 84 in 1850 his had been amputated I believe when he was about 4 years old & the regrowth was much larger & more clumsy than that on my son's hand— no other instances in the family are known but there was a curious legend in the family of a miller who was to be born with 3 thumbs who w^d hold a king's horse up to his knees in blood. My grandfather had a mill & 3 thumbs there the similitude ends!—

   …

   Will you kindly forward my letter & sketch to Mr. Darwin & tell him I shall be most happy to answer any other questions in my power, tho' I think I have given all the particulars I can

   With kind regards to Sir John | believe me | Y^{rs} very truly | Alice Lane Fox

Darwin evidently replied directly to Lane Fox, for she wrote again:

Uplands | Guildford
Aug. 3ᵈ./75

Dear Sir

I am very glad to be able to give you what ever further information in my power on the subject you request, but it is rather difficult at this distance of time to be perfectly certain of all the details. As regards the first point—the congelation was simply to deaden the pain & the thumb was cut off with scissors— the wound was a long time healing & I cannot distinctly remember how far there was any prominence left or when it began to grow—

I only remember we were disappointed at its not having been done more effectually & thinking another operation wᵈ. be necessary. I do not think there is any prominence of bone beneath the nail If at any time you were in London or anywhere convenient my son wᵈ. be most happy to show you his hand & let you judge for yourself—

I thought since I wrote to Lady Lubbock you wᵈ. be glad to know of another curious case of a school fellow of another son of ours at Charter House by name Smith who it appears has 6 digits on each foot as well as on his hands. He has to have his boots made very broad at the toe the extra digits are on the little finger— (He is nick-named hexagon) His parents are dead & his uncle is his guardian so it may be more difficult to get at the particulars, but his house tutor

Revᵈ. J. Evans
Charter House
Godalming

Whom I saw last night said he was sure the boy wᵈ. not mind giving you any information on the subject he cᵈ. & if you enclose a letter to his address I have no doubt he will do what he can in helping you to get the information. He said he had not seen his feet himself but going into the boy's dormitory the other night he heard Smith exclaim "by Jove its growing again" which gave him the impression he must have had another operation I shᵈ. like much to know to which case you refer in which yʳ accuracy has been attacked In "the Descent of Man" there are no special cases of supernumerary digits given I think?

believe me | yʳˢ. truly | Alice Lane Fox

At the same time, Darwin was writing on the subject to Annie Dowie, Robert Chambers's daughter. Robert Chambers was suspected by many, including Darwin, to be the anonymous author of *Vestiges of the natural history of creation*, published in 1844, which argued that scientific laws governed not only the development of higher life forms but the origin of life itself. Chambers himself had six digits on each hand and foot, and he had given Darwin the case of a child with an inherited extra digit that regrew when amputated; Darwin reported it in *Variation*. The child was Dowie's sister, Alice Chambers. In reply to Darwin's enquiries, Dowie sent a copy of notes in her father's diary, along with a letter that has not been found. Its content may be guessed from Darwin's answer:

Down, | Beckenham, Kent. | Railway Station | Orpington. S.E.R.
Aug. 1st

My dear Mrs Dowie

I thank you most warmly for all the trouble you have so kindly taken, & for your two very pleasant letters. You flatter me (though this is not the proper word) so delightfully about my old friends, the tendrils [*i.e. Darwin's book on climbing plants*], that I must send you my book just published about new friends, "Insectivorous Plants"; though there are only bits here & there which readable.——

I sh^d like very much to go to Bristol & have long wished to attend one more Brit. Assoc^n., but I cannot stand so much excitement & talk.

Now for business,——nothing can be clearer or fuller than your account; but I am much perplexed what to conclude, as Paget thinks the interest or importance of the case largely depends on the certainty of the removal of the base of the bone. He has known an amputated stump of the humerus to grow a little, but thinks this very different from a quite new bone being reformed.——. I hope that you will not think I have acted badly, when I say that to give him confidence I told him it was your Father who gave me the information & that Prof. Syme was the operator. I told him not to mention your Father's name to anyone & he says he has not done so; but has applied to Mr Annandale (Syme's assistant) who applied to his Sister to know whether Syme had ever mentioned *any* such case; but he never had to either.

Under these circumstances, & as Pagets excellent judgment w^d be of greatest value to me, I hope I may lay an *abstract* of your Fathers diary & of the facts which you mention, before him,——again cautioning him not to mention names to anyone. I feel bound in honour either to strike out whole case & confess to an error, or to substantiate my statement by details & by the judgment of an experienced physiologist & surgeon, like Paget.

Now there is one other thing, could you persuade your sister to make a tracing of her hand, the palm being placed quite flat on the paper with the pencil held vertically all the time. This would show form of the protuberance & its size. You would of course say that this tracing will be considered as strictly confidential.

Believe me that I feel deeply grateful to you, for to be indirectly accused of perverting the truth is the most painful acusation which can be made against me.

I remain dear Mrs Dowie | Yours truly obliged | Charles Darwin

P.S If your sister consents to oblige, perhaps it w^d be best to give two tracings, one with palm flat on paper, & the other with the back of the hand flat on paper.

Wetstones | West Kirby | Birkenhead.
10th August.

Dear Mr Darwin—

Very many thanks for your kind acknowledgment received yesterday. I have been examining my sisters hand again and can distinctly feel & see the fifth knuckle though it is smaller than the fourth.

I hope you will have an opportunity of examining it yourself some day. She tells me when she was 15—Mr Syme again examined her right hand and agreed with Papa as to the regrowth—but did not advise any fresh operation as it would require such deep cutting out, that it would most likely injure by stiffening her little finger. I am glad nothing more was done—as she plays most beautifully on the piano—and might have had great difficulty in doing so after such a serious operation.

Hoping to have the great pleasure of coming to see you some day at Down, and with very kind regards and much esteem

I remain, | dear Mr Darwin | Yours most sincerely | Annie Dowie.

Close to the end of his life, Darwin heard from Emily Fairbanks Talbot, an American living in Boston. Talbot had been a schoolteacher from the age of 16. Her enthusiasm for educational reform derived partly from her experience as a teacher, and partly from a desire for her daughters to have access to the best education possible. She tried and failed to enrol them in a number of male schools so that they could apply to Boston University. In 1877, Talbot established the Latin School for Girls, which offered college-preparatory classes equivalent to those of the Boston Latin School. In 1881, she helped found the Association of Collegiate Alumnae to further opportunities for female college graduates. She also organised the Massachusetts Society for the University Education of Women. She was interested in child development, and, although her letter to Darwin is missing, we have his reply to her. Darwin's suggestion of looking into whether education affected inherited traits, though here applied to race, was also relevant to his view of female intelligence, since he seemed to have thought it possible that the superior education and wider experience that had historically been offered to boys might have led to their inheriting a greater aptitude for learning.

Down, | Beckenham, Kent. | (Railway Station | Orpington. S.E.R.)
July 19th 1881

Dear Madam

In response to your wish I have much pleasure in expressing the interest which I feel in your proposed investigation on the mental & bodily development of infants.— Very little is at present accurately known on this subject, & I believe that isolated observations will add but little to our knowledge, whereas tabulated results from a very large number of observations systematically made, would probably throw much light on the sequence & period of development of the several faculties. This knowledge would probably give a foundation for some improvement in our education of young children, & would show us whether the same system ought to be followed in all cases.

I will venture to specify a few points of enquiry which, as it seems to me, possess some scientific interest. For instance does the education of the parents influence the mental powers of their children at any age, either at a very early or somewhat more advanced stage? This could perhaps be learnt by school-masters or mistresses, if a large number of children were first classed according to age & their mental attainments, & afterwards in accordance with the education of their parents, as far as this could be discovered. As observation is one of the earliest faculties developed in young children, & as this power would probably be exercised in an equal degree by the children of educated & uneducated persons, it seems not improbable that any transmitted effect from education would be displayed only at a somewhat advanced age. It would be desirable to test statistically in a similar manner the truth of the often repeated statement that coloured children at first learn as quickly as white children, but that they afterwards fall off in progress. If it could be proved that education acts not only on the individual, but by transmission on the race, this would be a great encouragement to all working on this all-important subject.

It is well known that children sometimes exhibit at a very early age strong special tastes, for which no cause can be assigned, although occasionally they may be accounted for by reversion to the taste or occupation of some progenitor; & it would be interesting to learn how far such early taste are persistent & influence the future career of the individual. In some instances such tastes die away without apparently leaving any after effect; but it would be adviseable to know how far this is commonly the case, as we should then know whether it was important to direct, as far as this is possible, the early tastes of our children. It may be more beneficial that a child should follow energetically some pursuit, of however trifling a nature, & thus acquire perseverance, than that he sh^d be turned from it, because of no future advantage to him.

I will mention one other small point of enquiry in relation to very young children, which may possibly prove important with respect to the origin of language; but it could be investigated only by persons possessing an accurate musical ear. Children even before they can articulate express some of their feelings & desires by noises uttered in different notes. For instance they make an interrogative noise, & others of assent & dissent in different tones, & it would, I think, be worth while to ascertain whether there is any uniformity in different children in the pitch of their voices under various frames of mind.

I fear that this letter can be of no use to you, but it will serve to show my sympathy & good wishes in your researches.

I beg leave to remain | Dear Madam | Yours faithfully | Charles Darwin

To | Mrs Emily Talbot.—

9 Editors

Editing provided a useful entry point into the world of science for women, even though this work was not usually acknowledged in print. For instance, Arabella Burton Buckley worked as Charles Lyell's secretary from 1864 (when she was 23) until his death in 1875. According to an article in a short-lived Liverpool journal, *Research: a monthly illustrated journal of science*, she was taken on as an amanuensis because of the clearness of her handwriting, but she became a secretary and indispensable literary assistant. She 'not merely wrote Sir Charles's letters after his instructions, and ordered his very extensive scientific correspondence, but when a new edition of any of his works was called for she drew illustrations, re-cast passages, made *précis* of information in new works, corrected the proofs, and compiled tables and indexes' (*Research*, 1 February 1889, p. 130). After Lyell's death, Buckley went on to have a distinguished career as a scientific author (see chapter 10) and lecturer.

In default of paid assistants, it was common for men of science to ask female friends and relatives to help. Sometimes help came unasked: Darwin's first editors were his sisters. Caroline Darwin, who was nine years older than him, educated him before he went to school in Shrewsbury at the age of 8. 'I doubt whether this plan answered,' wrote Darwin later; 'she was too zealous in trying to improve me' ('Recollections', p. 356). In this letter written to Darwin on the *Beagle* in 1833, however, Caroline is very guarded in her criticism:

October 28th—

My dear Charles—

I have been reading with the greatest interest your journal & I found it **very** entertaining & interesting, your writing at the time gives such reality to your descriptions & brings every little incident before one with a force that no after account could do. I am very doubtful whether it is not *pert* in me to criticize, using merely my own judgment, for no one else of the family have yet read this last part—but I *will* say just what I think—I mean as to your style. I thought in the first part (of this last journal) that you had, probably from reading so much of Humboldt, got his phraseology & occasionly made use of the kind of flowery french expressions which he uses, instead of your own simple straight forward & far more agreeable style. I have no doubt you have without perceiving it got to embody your ideas in his poetical language & from his being a foreigner it does not sound unnatural in him— Remember, this criticism only applies to parts of your journal, the greatest part I liked exceedingly & could find no fault, & all of it I had the greatest pleasure in reading—

...

dearest Charles good bye. Y^rs affly C D.

Susan Darwin, only six years older than Darwin, was more forthright (her nickname was Granny). She sent Darwin birthday greetings on 12 February 1834, together with some corrections:

My Sisters have told you how very much we enjoyed your Journal and what a nice amusing book of travels it w^d make if printed, but there is one part of your Journal as your Granny I shall take in hand namely several little errors in orthography of which I shall send you a list that you may profit by my lectures tho' the world is between us.— so here goes.—

wrong	right according to sense.
loose. lan*s*cape. hig*est*	lose. landscape. highest.
profil. cann*a*bal	profile. cannibal. peaceable
peac*i*ble. quarre*ll*	quarrel.—

I daresay these errors are the effect of haste, but as your Granny it is my duty to point them out.—

Darwin took the advice in fairly good part, responding to Catherine Darwin, his younger sister, on 20 July 1834: 'Thank Granny for her purse & tell her I plead guilty to some of her [*two words obliterated*], but the others are certainly only accidental errors— Moreover I am much obliged for Carolines criticisms (see how good I am becoming!) they are perfectly just, I even felt aware of the faults she points out, when writing my journal' (*Correspondence* vol. 1).

He also later relied to some extent on readers among his friends. The earliest of these that we know of was Georgina Tollet, a close friend of the Wedgwoods and Darwins in Shropshire. Tollet had her right arm amputated when she was eighteen owing to an abscess, and had to learn to write and draw with her left hand. She was the author of the posthumously and anonymously published *Country conversations* (1886). The preface gives a classic portrait of the self-effacing female editor:

The writer had a singularly accurate memory, a sense of quiet humour, and keen powers of observation; but of the faculty which creates she had no share. Her sole object was to preserve the exact expressions of those whose histories of themselves and of their affairs she had found so interesting. She scrupulously avoided making any additions or changes, though she sometimes omitted trifling details, and recorded as little as possible of her own share in the dialogue.

The conversations are often extremely funny and amongst other things give a vivid and unexpected account of the importance of cheese-making in the domestic economy and marriage markets of rural Shropshire, and of the social power of cheese-making women.

In 1859, she read the manuscript of *Origin of species*. Darwin wrote to his publisher, John Murray, to ask him to send it to her.

> Down Bromley Kent
> April. 5th.—
>
> My dear Sir
>
> I send by this Post Title (with some remarks on separate page) & 3 first Ch⁵. If you have patience to read all Ch. I, I honestly think you will have fair notion of interest of whole book.— It may be conceit, but I believe the subject will interest the public & I am sure that the views are original.— …
>
> As soon as you have done with M.S, please to send it by *careful messenger* & *plainly directed* to
>
>> Miss G. Tollett
>> 14. Queen Anne St
>> Cavendish Sq^re
>
> This lady being excellent judge of style is going to look out for errors for me.

Emma Darwin had lunched with the Tolletts during her recent trip to London, from 1 to 4 April 1859 (Emma Darwin's diary (DAR 242)). The arrangement may have been made then.

Joseph Dalton Hooker's wife, Frances Harriet Hooker, also read the manuscript in 1859, although we don't know whether she was asked to by Darwin or by her husband.

> Down Bromley Kent
> May 11
>
> My dear Hooker
>
> Thank you for telling me about obscurity of style. But on my life no nigger with lash over him could have worked harder at clearness than I have. But the very difficulty to me, by itself leads to probability that I fail. Yet one lady who has read all my M.S. has found only 2 or 3 obscure sentence. But Mrs Hooker having so found it, makes me tremble.— I will do my best in proofs. You are a good man to take trouble to write about it—

According to Francis Darwin, Emma read the page-proofs of Darwin's books, 'chiefly for misprints and to criticise punctuation'; and Darwin used to dispute with her about commas especially (DAR 140.3: 145). We know she was also deeply concerned about the ultimate tendency of his work, but this 1871 letter to Alfred Russel Wallace gives an indication of her very practical approach to Darwin's scientific career. Darwin was in two minds about whether to have a paper in support of his views by Chauncey Wright, an American philosopher, reprinted in England. Wright's paper involved some fairly abstruse philosophical reasoning, and Wallace had advised against republishing it.

> *Down, | Beckenham, Kent.*
> July 12th

My dear Wallace

Very many thanks. As soon as I read your letter I determined not to print the paper, notwithstanding my eldest daughter, who is a very good critic, thought it so interesting as to be worth reprinting. Then my wife came in, & said "I do not much care much about these things & shall therefore be a good judge whether it is very dull". So I will leave my decision open for a day or two.

Darwin did have the paper reprinted, although we don't know whether that was on Emma's advice or not. Evidently Emma pitched herself as less of a scientific connoisseur than Henrietta.

Until her marriage in 1871, Henrietta, Darwin's eldest surviving daughter, was his chief assistant in writing and publishing. After her marriage, she still worked on his books, although Darwin took on his son Francis as a scientific secretary, and asked his sons George and William to help with checking proof-sheets.

Henrietta was born in 1843, the Darwins' fourth child after William, Annie, and Mary, who lived less than a month. The much-beloved Annie died in 1851. Six more children were born after Henrietta. Henrietta herself suffered the family ill health: she was inclined to poke fun at herself and the others for being 'sickies'. One story has it that as a child she was told by a doctor that she might have breakfast in bed, and never got up for breakfast again. Undoubtedly she grew up 'bookish', reading both her fathers' works and those of his scientific colleagues. In 1862, she was a test reader for Darwin's work on the fertilisation of orchids. Emma wrote to William: 'Papa is correcting the press of the orchis's & he gets Hen. to read it over to see whether she understands it & she finds it not easy' (DAR 219.1: 49).

Henrietta may have struggled with the technical terminology in the orchids book, but she was clearly no fool. In 1863, Darwin passed on some of her criticisms of Thomas Henry Huxley's *Lectures to working men* to the author. 'Please to say to Miss Henrietta Minor Rhadamanthus Darwin that I plead guilty to the justice of both criticisms & throw myself on the mercy

Henrietta Darwin. DAR 225: 55.
By permission of the Syndics of Cambridge University Library.

of the Court', Huxley responded. (Rhadamanthus is one of the judges in Hades in Greek mythology.) The Huxley and Darwin families, both well endowed with children, were very close, and Huxley was no doubt used to hearing Henrietta's views at first hand.

Henrietta probably did more work than any other reader on Darwin's books; she undertook detailed editing of his major works, and since she sometimes did the work while travelling, some letters on the subject survive. The first work that we know she edited, in 1867, is *Variation of animals and plants under domestication.*

July 26th

My dear Etty.—

You are a very good girl to wish for remaining slips of present chapter, but they are enormously altered & 10 folio pages of MS added, & the slips themselves have had to be cut into pieces & rearranged, so I will not send them.

But for the future I shall be only too glad for you to see the slips, as well as Revises.— I will either keep, according to quantity finished, the whole of present chapter till your return, or send part to you.—

All your remarks, criticisms doubts & corrections are excellent, excellent, excellent

Yours affect^{ly} | C. D.

Henrietta spent the second half of July 1867 visiting Devon and Cornwall; in a letter to her brother George she mentioned having '11 pages of proof to do' (letter to G. H. Darwin, 23 July [1867], DAR 245: 280). Darwin paid Henrietta £20 for correcting the proofs.

A couple of brief references in the family letters reveal that Henrietta translated Federico Delpino's 'Brief remarks on the biology and genealogy of the Marantaceae' from Italian; the translations appeared in the weekly journal *Scientific Opinion* in two parts, on 2 and 9 February 1870. Emma wrote to her, 'Your paper looks v. knowing in Scientific Opinion', and on 11 February, 'F [*father*] descries me to tell you that your 2nd part is come out in Scientific Op. & reads v. smooth & well' (DAR 219.2: 72, 73).

Henrietta also edited *Descent of man*, by which time she was clearly a valued reader. Darwin wrote to her on 8 February 1870.

My dear H.

Please read the Ch. first **right through** without a pencil in your hand, that you may judge of general scheme; as, also, I particularly wish to know whether parts are extra tedious; but remember that M.S is always *much* more tedious than print.— The object of Ch. is simply comparison of mind in men & animals: in the next chapt. I discuss progress of morals &c.— Some sentences are at back of Page marked thus @.—

I do not send foot-notes, as I have no copy & they are almost wholly mere authorities.— After reading once right through, the more time you can give up for deep criticism or corrections of style, the more grateful I shall be.— Please make any long corrections on separate slips of paper, leaving narrow blank edge, & pin them to margin of each sheet, so that I can turn each back, & read whilst still attached to its proper page.— This will save me a world of troubles Heaven only knows what you will think of the whole, for I cannot conjecture.— You are a very good girl indeed to undertake the job.—

Your affect Father | C. Darwin

(I suspect that here & there style will want a good deal of improvement, though I hope greater part fair.—)

(I fear parts are too like a Sermon: who w^d ever have thought that I sh^d turn parson?)

Henrietta had been reading parts of the manuscript of *Descent* at least since she left England for France and Italy in January 1870 (letter from Emma Darwin to H. E. Darwin, [18 January 1870] (DAR 219.9: 69)). With this letter Darwin had presumably sent the second chapter on the comparison of the mental powers of humans and animals (*Descent* 1: 70–106). Henrietta replied:

Dear F.

Thanks for your note abt the M.S.S. I will obey all your instructions—& I shall be extremely interested in reading it—so you needn't thank me for it—& here what time I spend in my own room is so very undisturbed it goes much further. When I know no human being will come after me & lay out my plans for the day & stick to them. Certainly to have you turned Parson will be a change— I expect I shall want it enlarging not contracting—cos I think *you* think an apology is wanting for writing abt any thing so unimportant as the mind of man!

Darwin was delighted with Henrietta's work, although he couldn't resist a fatherly dig at her handwriting.

Spring 1870

My dear Hen.

I have worked through, (& it is hard work) half of the 2^d Chapter on mind, & your corrections & suggestions are *excellent*. I have adopted the greater number, & I am sure that they are very great improvements.— Some of the transpositions are most just. You have done me real good service; but by Jove how hard you must have worked & how thoroughily you have mastered my M.S. I am pleased with this Chapter now that it comes fresh to me.—

Your affectionate, admiring & obedient | Father, C. D.

All is as clear as daylight— your plan of putting corrections saves me a world of trouble, by just as much as it must have caused you.— N.B. you **can** write, I see, a perfectly clear hand, as in *all* the corrections.—

As before, Henrietta was thanked and paid for the work, although the payment was politely offered as a memorial or souvenir rather than as wages.

Down
March 20. 1871

My dear Henrietta,

I do not know whether you have been told that Murray reprinted 2000, making the edition 4500, & I shall receive 1470£ for it. That is a fine big [*sum*]. The corrections were 128£!! Altogether the book, I think, as yet has been very successful, & I have been hardly at all abused. Several reviewers speak of the lucid vigorous style etc.— Now I know how much I owe to you in this respect, which includes arrangement, not to mention still more important aids in the reasoning. Therefore I wish to give you some little memorial costing about 25 or 50£, to keep in memory of the book, over which you took such immense trouble. I have consulted Mamma, but we cannot think what you would like, & she with her accustomed wisdom advised me to lay the case before you & let you decide how you like—

I have been greatly interested by the second article in the Spectator & by Wallace's long *article* in the Academy— I see I have had no influence on him, & his Review has had hardly any on me—

We go to London on April 1st for a few days, in order that I may visit & consult Rejlander about Photographs on Expression— I think I shall make an interesting little vol. on the subject.— By the way, I have had hardly any letters about 'the Descent' worth keeping for you, except one from a Welshman abusing me as an old ape with a hairy face & thick skull. We shall be heartily glad to see you home again.

Goodbye my very dear coadjutor & fellow-labourer | Your affec^ate. father. Ch. Darwin.

Sea Grove | Bournemouth
March 21st

My dear Father

Thank you very much for your letter this morning— I am v. glad old Murray got up his courage for the extra 500 so as to put off the evil day of a 2nd. edit— £1470 is a splendid sum & now I hope you will be easy abt Murray's gains in spite of the 128£— I think some of those great novelists only get £3000—I can't remember who—&

to think of that kind of book bringing in nearly as m. as ½ a novel is wonderful.

What you say abt my helping has pleased me v. m— The pleasure of doing it rewards me for any trouble I can take over & over again— but to have a say so much, & to feel that at any rate *you* think I can help you so really, is very sweet to me— The memorial you propose will be very precious to me. I can't think all of a sudden what I sh.ᵈ like to have that will be appropriate & lasting— I want it to be something that will seem fitting in the nature of things—& something that I shall like for always— & so deep reflexion is required. It is very good of u to think of it—

...

What a funny Welchman your man must be for if u come from a hairy ape u are likely to be one & tis an argument for you— I haven't seen the 2.ⁿᵈ Spec. I'm rather surprised u've had no effect on Wallace. It seems to me his mind can't be so clear as u used to think it for I'm *sure* u are right. I've been rereading the old Physical basis [*Thomas Henry Huxley's essay 'On the physical basis of life'*] & have at last worked out on paper my dissatisfaction with it. If he wasn't such a busy man & the article hadn't been worked threadbare & I was likely to see him, I sh.ᵈ like to see how neatly he w.ᵈ smash me into a cocked hat— ...

Thank you again dear Father— your most affect | HED

In 1871, Henrietta married, and was in theory not so available as an editor, but in 1872, she was working again on *Expression of the emotions.*

Down.
July 25ᵗʰ 1872

My dearest H.

What a deal of pains you have taken over the chapt.— I am quite sorry that you sh.ᵈ have had the trouble of writing out cleanly your corrections, though you thus saved me much trouble. It was, however, a tough job considering all your alterations, almost everyone of which has been accepted & all are good.— I struck out the long par. about which I asked you; though I did so at last with some regret.— When in doubt do *not* take your trick is a golden rule, I believe, in writing.— I agree to what you say about latter pars. in Chapt. & I have partly accepted your alterations. In the last Par. I cut the Gordion Knot by leaving out all about the philosophy of language. It ends rather flat, & flat it must remain.

If you have nothing to say, say it, is not a golden rule in writing.

Very many thanks, I hope I have not killed you. I know that I am half-killed myself.—

Yours affect., | C. Darwin

In 1874, Henrietta was working on the proof-sheets of the second edition of *Coral reefs*, evidently checking that Darwin's manuscript corrections, and her own, had been made. At the same time, Darwin, with George's help, was working on a second edition of *Descent*. This letter was written on 21 March 1874. The text is from a copy; the copyist left blanks when unable to read Darwin's handwriting.

21.

My dear Etty

You are a good dear girl to take so sweetly all the horrid bother of correction. You ⸻ me much. I have taken all your corrections, except one small one. I am glad you have not much more, though parts are dreadfully written.— I was just recovering from my longest illness. When I corrected my M. S & suppose I had not strength enough to correct, as I cannot otherwise understand my horribly bad writing.

Do not think of comparing old & new text, *except references* as pages often get printed wrong.— You need not compare my M.S. additions as the sense will show whether they are correct.— It would save me some trouble, if you were to ⸻ your corrections, when not too long, on margin of sheets in pencil & then your mother could ink them in, & never mind giving your reasons as this must cost you more trouble. Again I say you are a good dear girl & your husband is so good a man, that I do not believe that he will grudge me this assistance. The Descent half kills me.—

Your affect Father | C. Darwin

Darwin's confidence in Henrietta's judgement and a certain lack of perfectionism on his own part are evident. When he asked George to help him with the second edition of *Descent,* he gave him a warning: 'I will give you 2 cautions, viz. not to alter the strength of my expressions & 2^{ndly} not to improve my style too much, as I deliberately think that it is best for each man to retain his own style, & that rather rugged sentences do not signify, if they are perfectly clear (i.e. as unlike Snow's as possible.)' (Snow was Frances Julia Wedgwood, Darwin's niece.)

Darwin occasionally heard from other 'family' editors, such as Marion Bell (see chapter 4), and in 1871 received a letter from the novelist and poet Rosa Mackenzie Kettle, who was editing a friend's work after his death, having evidently been his literary confidante for years. Little is known of Kettle, although she was a fairly well known novelist in her day, meriting a brief obituary in the *Publisher's Circular*; she was born Mary Rosa Stuart Kettle in 1817 in Overseal, Leicestershire, and but she and her two sisters adopted the name Mackenzie Kettle instead of Stuart Kettle. Her sister Clara, an artist, illustrated her books (*Art-Journal*, 1 April 1872, p. 128), and wrote on art and religion. Rosa, Clara, and a third sister, Harriet, appear in Poole in Devon, all unmarried and describing themselves as fundholders, in the 1871 census.

Heathside near Poole
Friday March 10

Sir/

Will you forgive me for troubling you with the request to be allowed to make use of some sentences in letters of yours to the late Charles Boner author of "Transylvania" "Forest creatures" and other works? I have been occupied for many months with the arrangement of his papers and I have made a collection of Memorials of his varied and most interesting career at the request & with the assistance of his daughter and sister.

Mr. Boner consulted me on literary matters and when he died no one but myself was acquainted with his wishes & instructions respecting his last work— While preparing it for the press it struck me that some notice of its author might be added and so many clever men have aided & contributed that my tribute to his memory has become much more important than I at first expected.

Among many private letters & papers sent to me from Munich by Madame Horschelt and Miss Boner [*Charles Boner's daughter and sister*] I find some letters of yours which contain such a true appreciation of my dear friends works and powers of observation that I would willingly if you have no objection make a few extracts from them. May I use your name as it will increase the value of your flattering comments? If you have any letters of Charles Boner's likely to be interesting either whole or in portion to the public it would be a great favour to me and to his daughter if you would lend them to me. I will promise to return them if you wish it and you may rely on my discretion not to publish any passage which could in any respect be objectionable.

I must add that all I do in this matter is out of respect for M̲ʳ̲ Boner's talents and affection for his memory as I shall derive no pecuniary benefits from the work. For the accuracy of what I have now said as I am a perfect stranger to you I can refer you to M̲ᶜ̲ Horschelt, M̲ʳ̲ Boner's daughter, who resides at 5 Louisa Strasse Munich. She has entrusted me with all her father's most private papers & letters and relies on me implicitly.

Believe me to remain | with much respect | yours very faithfully | Rosa Mackenzie Kettle

An easier reference perhaps would be Edward Wilberforce Esqʳᵉ

4 Harcourt Buildings
Inner Temple

Please to address me as Miss Rosa M. Kettle

an early reply will oblige me as I am preparing the proof sheets for publication by M̲ʳ̲ Bentley after Easter

Darwin eventually had offers of the assistance of women he didn't know. The surgeon Lawson Tait wrote to Darwin on 5 June 1875:

If I can ever at any time relieve you of any drudgery in correcting proof, pray let me know. My wife does all mine & she is the sharpest hand at it I have ever met with. She will be delighted to ease the labours of one whose writings she knows so well

Sybil Anne Tait was a member, like her husband, of the Birmingham Natural History Society and the Birmingham Philosophical Society. Little is known of her, except that she took an interest in Tait's work and was a good proofreader. Tait himself was opposed to vivisection and cautiously in favour of women's rights (those two causes often went together); his women's hospital in Birmingham had female surgeons and he himself had a female anaesthetist. It's difficult not to be sceptical of Tait's generosity in offering his wife's services and of his confidence in her delight in drudgery: but at the same time, it's clear that editorial work did offer women an unobtrusive way into the scientific community, a moderate amount of status, and some intellectual nourishment.

10   Writers and critics

Writing, like being a governess, was a respectable profession for middle-class women who needed to earn a living, and could be a satisfying means of self-expression even for those who didn't. Victorian society had an insatiable desire for novels, journalism, and science. However, feeding that desire was rarely very profitable, and women who relied on writing alone for an income often had to apply to charitable organisations like the Royal Literary Fund, or petition for a civil-list pension. Women did write under their own names, although it was also common to use pseudonyms or initials; critics could be patronising to female authors. Some of the women in this chapter were closely involved with the feminist movement; most were acutely aware of the disadvantages they were under in not having had the education they would have chosen.

Eliza Meteyard's parents had been acquainted with Darwin's parents in Shrewsbury. She never married, but earned her living from writing. The option of earning a living as a governess was not open to her, as she was deaf. She was a council member of the radical and feminist Whittington Club, and a prolific author of novels and journalism, both under her own name and under the name Silverpen. Her interest in pottery, architecture, and design was bound up with the early arts and crafts movement. Most famously, she wrote a biography of Josiah Wedgwood, the celebrated potter. Hearing that she was engaged in this work, Darwin in 1863 sent her via her publisher letters between his paternal grandfather, Erasmus Darwin, and Josiah Wedgwood, his maternal grandfather. Meteyard's letters suggest a harried existence, with little of the financial security or practical assistance that Darwin enjoyed. The heroine of her semi-autobiographical novel *Struggles for fame*, faced with a proposal of marriage, announces, 'The woman who wishes to excel in literature must be alone from the cradle to the grave.' Instead, Meteyard relied on a network of friends.

In 1865, Meteyard sent Darwin a copy of the first volume of her biography of Wedgwood.

Wildwood | North End | Hampstead. N.W.
April 25. 1865

Dear Sir

Permit me to offer you, in the hope of your acceptance, the 1st Vol of the 'Life of Wedgwood'. I regret I have to send the work in this partial form, but the envy and machinations of two unscrupulous men rendered the step imperative. The other volume will appear as soon as possible.

I lay the book before you with a trembling hand, and with sincere and great humility of spirit. I know what the subject requires—and I would that I had the highest human ability to do justice to my conception of it. Yet where I fail—my publishers splendid justice to the work—must be some compensation—and the veneration I have for the names of Darwin and Wedgwood, and the liberal point of view from which I judge them and their opinions—must make up—as I hope it will in your kindly eyes—for deficiencies of various kinds. A bigot—whether social, political, or religious, *could not* estimate the character of two such men. Men far greater than their generation.

I am using your valuable letters as I go on. I shall not have space I fear to do full justice to the more scientific aspect of Wedgwood's labours and bent of mind in *this* work—but—as I am going to carry on the details of these and other subjects, through a work I shall call 'Thomas Wedgwood and his Contemporaries' I shall be able to find a still more fitting place for many interesting though severer truths. As soon as I have finished the Life of Josiah Wedgwood, I will make what further extracts I need from your valuable letters—and carefully return them. Meanwhile they are in excellent keeping.

I trust your health has improved, for missing that—we miss almost the best thing we hold in life. Do not please trouble yourself to write— a line from Mrs Darwin just to say the book has reached you safely—is all that is required.

Dear Sir | With deep respect & grateful obligation | your's faithfully | Eliza Meteyard

A few months later Darwin asked to have one or more of the letters returned to him; the editor of the *Autographic Mirror* wanted a sample of Erasmus Darwin's handwriting (*Correspondence* vol. 13, letter to Eliza Meteyard, 16 November [1865]). Meteyard replied:

> Wildwood | North End | Hampstead. NW.
> Nov: 17. 1865

Dear Sir,

I send you nineteen of the letters you so kindly lent me: and hope amongst the number, one may be found suitable for the pages of the Autographic Mirror. If not, and Mrs Darwin will kindly inform me, I will send the rest, although dipping into them for the Wedgwood as I go on.

I heartily apologize for retaining them so long, but I conscientiously feel that no one can reverence them more, or take greater care; and this perhaps makes me take unwarrantable liberties as to time. But your goodness I am sure pardons me. The circumstances under which my book has had to be written has much harrassed me—and with hundreds of pages to copy, and no helping hand—I necessarily take liberty where I can. But as soon as the last page of Wedgwood is off to the printers—and we begin to print though slowly on Dec. 1— I will set to, at once, and make what notes I shall further require for the Life of Thomas Wedgwood—and then return them speedily—with very sincere thanks for your indulgence.

I sincerely hope your health is a little better than it was.

With compliments to Mrs Darwin, I am, | Dear Sir | With grateful respect | your's faithfully | Eliza Meteyard.

In 1869, Meteyard asked Darwin to support her application for a civil-list pension.

> wildwood. | Hampstead. N.W
> Feb. 19. 1869.

Dear Sir,

I received the memorial from Sir John Lubbock—quite safely—the evening before last. It contains his valuable signature as also your own—and that of Sir Charles Lyell. When you see or write to these gentlemen, will you offer to them my very earnest thanks for their true goodness. And will you please receive the same yourself—with the addition of my deep respect. Such names are of the highest value to me—not only for the intrinsic weight they bear—in relation to the matter in question—but to myself as a silent guarantee of opinion. Believe me truly and respectfully grateful to you.

The memorial is now with Mr Winter Jones at the British Museum—for his and others' signatures. But I am somewhat in a commotion—Mr Smiles having written urgently last night—for the memorial to be sent in to Mr Gladstone. There will therefore be only time to add the names of my kind friends—Mr Duffus Hardy and Mr Bennett Woodcroft. There are many names of great value— awaiting the memorial—as those of Mr Warren De La Rue—Mr Millais—and others—but there is now no time for application.

Believe me—with great respect, | your's obliged & faithfully | Eliza Meteyard

In 1874, Meteyard made another application.

> 5 Squires Mount. | Hampstead. N.W.
> April 20. 1874

Dear Sir,

I send you my Memorial to ask you to have the goodness to sign it *once more*. It failed—on being presented last autumn—simply for the reason that applicants were so numerous—but Lord Sandon who presented it—promised that if his party came into power, I might hope for better success. Accordingly my kind friend Mr Falke is at work again for me—& to the form of Memorial in his hands—he has obtained some excellent signatures.

I hope you will not think me intrusive or troublesome. My health has wholly given way owing to my life of constant anxiety & toil without respite of any kind—& the addition of £40 per annum to what I have at present—£60—would be the greatest possible boon.

I have now passing through the press—the 'Handbook for Collectors of Wedgwood-ware' of which I shall venture to send Mrs Darwin a copy as soon as it is out.

With great respect— | Yours faithfully | Eliza Meteyard.

5 Squires Mount. | Hampstead. N.W.
June 27. 1874

Dear Sir,

As you so kindly appended your valuable signature to my Memorial I am sure you will be pleased to hear that it has been successful, & that Her Majesty, on the recommendation of Mr Disraeli, has increased my Pension on the Civil List to £100 per annum for the rest of my life. This is a great boon—freeing my mind—as it will do from the constant anxiety of how to live—& enabling me to bend my attention to better work—than any yet done—among which—in a new edition—will I hope be a more perfected Memoir of your illustrious grandfather—Mr Wedgwood.

I have begun my MS book in relation to "the Darwins" & will send it in November with the 'Handbook'. The letters, so far as I have copied relate to the proof sheets of the lines on the Portland Vase. Dr Johnson of Shrewsbury has made a collection of letters & papers relative to—as also written by, Dr R. W. Darwin. Last year Dr Johnson offered them to me for literary use. I have none. Still they might—if copied—be serviceable to the future historian— so, as next month I may be in Shrewsbury—I think of taking loan of the papers & forwarding them to you to look over. Such material so easily perishes, or is lost sight of.

When at work last autumn on 'Memorials of Wedgwood' I was deeply struck with the rude vigour of the subject of one of 'Wedgwood & Bentleys' plaques—of which—with this is an autotype—which please retain. The humanization of the fauns & satyrs is most wonderful. Where Wedgwood derived this subject from I cannot say—though the treatment is that of the Renaissance. To me it appears that all forms of this class were primarily derived from living objects; & that the antique artists simply vitalized descriptions of half humanized forms handed down profoundly remote—yet most reliable tradition. The satyrs, the fauns, the pigmies are undoubtedly derivations of actual forms—just as in our own country as elsewhere—

the traditions of great worms, dragons, & other mighty reptiles & animals living in & creeping out of caves—point to the time, when man commenced his warfare with, & subdued, the last of the prehistoric fauna. Apropos to the fauns & satyrs, I came upon a remarkable passage in King's work on ancient gems. He quotes from Cæsar—& Cæsar, if I recollect, refers to the Phœnician voyages. The passage refers to faun-like forms seen on the heights of Gibraltar.

I hope you will not think me presumptuous in referring to these subjects—but they are those in which I take a profound interest. I have been a great reader of your books—& had I had culture sufficient I should be very scientific. From association with my brother William—dead many years—I derive these predilictions. Among his books—was one—which if you have never seen—you may like to look over—& as it is of no use to me—I shall be honoured if you will place it on your book-shelves. Verity was a wonderful man—but he died young—in Paris. He was an Englishman. [*Meteyard probably sent Robert Verity's Changes produced in the nervous system by civilization (1837).*]

Hoping you will pardon this long note. With compliments to Mrs Darwin. I am—with great respect | Yours obliged & truly | Eliza Meteyard.

In Mr Lewes's new work 'Problems of Life & Mind' is an argument very ably brought out—as to the non-necessity of "missing links"

No answer to this note is required, and if when Mrs Darwin sees or writes to Mr Hensleigh Wedgwood—she will thank him—from me—for his signature & will tell him of my success—I shall be very much obliged.

Lydia Becker, who wrote to Darwin about botany, did not earn a living from writing, but she did publish one book, *Botany for novices: a short outline of the natural system of classification of plants* (under the initials L.E.B.), as well as pamphlets and articles on women's suffrage and women's access to science. She also produced a manuscript on astronomy that was never published. *Botany for novices* is sixty pages long, illustrated, and aims to explain botanical terminology and plant structure in a straightforward manner, thus giving a reader an entry point for more advanced study. It accorded well with Becker's belief that women should be encouraged to study science as an indispensable enrichment of their lives.

Altham | Accrington
March 30th. 1864

Dear Sir

You were so indulgent to me when I troubled you with a communication last summer that I presume on your kindness so far as to take the liberty of offering for your acceptance the accompanying little book in the hope that you may look kindly on my endeavour to make plain by familiar language and illustration the general principles of the subject to which it relates. It is intended chiefly for young ladies but I trust this circumstance alone would not cause you to consider it beneath your notice, for it is precisely those who have attained the greatest eminence in the pursuit of science who might be expected to feel pleasure in the thought that others however far removed from them, should be led to share in some degree, the happiness which the study of nature is capable of affording. Therefore I working for those who have yet their alphabet to learn, venture to hope that my attempt may find favour from you—at the other end of the scale— who have done more than any other to arouse general interest in the science you love so well and who have made plain for future explorers— the path in which henceforward they must all proceed—

I remain dear Sir | yours respectfully | Lydia E. Becker

Like Eliza Meteyard, Mary Somerville made a living partly from writing and from a civil list pension, although unlike Meteyard she married twice and had a small inheritance from her first husband. She had a natural bent for mathematics, but received little encouragement from her parents in this or any other study other than needlework and reading the Bible. An uncle, however, encouraged her study of Latin, and once she was recognised as a gifted writer and interpreter of science, she was much fêted by her scientific contemporaries. The Royal Society of London commissioned a marble bust of her (although it could not make her a fellow), the Royal Irish Academy made her a member, and the Royal Astronomical Society made her an honorary member. She was more than forty years older than Lydia Becker, and got a start in scientific society at a time when it could be done to some extent through private social contacts. Perhaps this, and her status as an exception, accounts for her very different experience of the problem of membership (see chapter 14). The generational difference may also account for a apparent difference between Becker's and Somerville's experiences; Somerville resented the lack of formal education consequent upon her sex, but she was inspired to study mathematics seriously by problems set in a fashion magazine. In the early nineteenth century, it was not unusual for women's magazines, notably the *Ladies' Diary*, to tackle fairly advanced mathematics.

The very brief exchange of letters between Darwin and Somerville has the relaxed yet respectful tone one might expect between two recognised leaders in their fields. Via her friend Mary Lyell, Somerville asked to re-use some of Darwin's illustrations from his book on orchids in her last book, *On molecular science*. Darwin's reply was probably written on 19 October 1866.

Down
Friday

My dear Lady Lyell

I should be delighted & honoured by M⁵ Somerville's using any of the diagrams in my Orchid book. But it is more M⁵ Murray's affair than mine. If this note were shewn to him I have no doubt he would give permission & do what is necessary.

Pray believe me yours sincerely | Ch. Darwin

Somerville sent her thanks from Italy:

> La Spezia Piemonte
> 30th Octr 1866

Dear Mr Darwin

I beg of you to accept my very sincere thanks for your kindness in granting me permission to make use of the illustrations in your admirable work on the Orchids, which I have taken the liberty to refer to very extensively in a sketch on the present state of microscopic science which I have just finished

Believe me | very gratefully yours | Mary Somerville

In 1869, Somerville sent Darwin a copy of the book.

> *Down.* | *Bromley.* | *Kent. S.E.*
> Jan 21st.

Dear Madam

I beg leave to thank you very sincerely for your great kindness in sending me a copy of your beautiful work, received yesterday, on Molecular & Microscopic Science, and for the honour you have thus conferred on me.

I have not yet had time to do more than turn over the pages, & to read the part about Orchids, in which you give an excellent summary of the subject.

With my best thanks & the most sincere respect, | I beg permission to remain | dear Madam, | yours truly obliged | Charles Darwin

Frances Power Cobbe was a friend of Somerville's, and a campaigning journalist in the anti-vivisection movement. She had been fairly expensively educated for two years, and bitterly resented it as an interruption to her home-schooling; her father's death left her rather hard-up, although she managed to travel in Europe, Egypt, Lebanon, Palestine, and Syria before returning to England. She wrote on morality, philosophy, theology, feminism, and animal behaviour. She met Darwin at his brother Erasmus's house, and consolidated the relationship when Darwin stayed near her and her companion, Mary Lloyd, on a summer holiday in Wales.

With this letter, probably from 1870, Cobbe sent a copy of her review of Francis Galton's *Hereditary genius* and Prosper Despine's *Psychologie naturelle* from the *Theological Review*.

Frances Power Cobbe. Wellcome Library, London (CC BY 4.0).
L0010481.

March 28th

Dear M[r] Darwin,

I feel very proud of having inveigled you into "looking through" Kant—Though I cannot quite say, like one of his disciples, "God said: Let there be Light, & there was—the Kantian Philosophy" yet I have retained for these twenty years a most lively sense of gratitude to him for helping me to find (or think I found) a stepping stone or two in the Slough of Despond

I more than suspect you of a smile in your beard when you write of him as "a great philosopher looking exclusive into his own mind"— But surely may I not argue that, after all, his mind, & that of another philosopher I could name are things not wholly undeserving of attention,— phenomena quite as much needing to be studied & accounted for, say, as even our beloved dogs? We poor humble learners who would fain be the most docile of your scholars, see one of you driving complacently down the "high priori road", & the other with infinite skill progressing on the solid causeway of material facts— But are you never going to unite your lines of thought & let us see how metaphysics & physics form one great philosophy?—

Pray forgive dear M[r] Darwin, my infinite impudence! Though I attended on Saturday a most successful Woman's Rights Meeting I am of opinion that our Ancient privilege of talking nonsense even to those we most deeply honour, is one not to be parted with on any terms!—

I enclose a little notice embodying what I thought the point of those "Cut pages" of Despine— Do not think of acknowledging this or returning *that*—

Most truly your's | Frances Power Cobbe

Emily Jane Pfeiffer was a Welsh-born poet; her interests in poetry, painting, and embroidery were fostered by her father. She married a German tea merchant resident in London in 1850. She contributed essays to the *Cornhill Magazine* and the *Contemporary Review* about the position of women in society, and her poetry often explored the theme of female disempowerment.

A critic in *The Times*, 8 January 1874, p. 4, used his generally favourable review of Pfeiffer's *Gerard's monument, and other poems* as the starting point for some remarks on female poets and women in general. A woman who could write poetry of the first order was the rarest phenomenon in the universe, the reviewer asserted. Poetry was the 'sum and substance of human life', and human life was 'more masculine than feminine': the work of the world was done by men. Women had not been prevented from becoming poets as they had been prevented from becoming soldiers or members of Parliament: they had tried and they had failed. However, Pfeiffer's work showed that it was quite possible for a woman to write verse that would be agreeable even to palates that scarcely cared to 'quench their thirst with anything less than the nectar of the gods'.

Pfeiffer wrote a response which is bound in the back of some copies of *Gerard's monument* together with a transcript of this marvellously condescending review. 'That we can be speculated about now in this advanced stage of the world's history more as if we were some extinct species than beings who have stood side by side with man from the beginning, is itself a striking result of that tyranny of circumstances which has retarded female developement. ... Every authoritative announcement of woman's inherent disqualification for the highest labours of the mind retards the issue which time has still to resolve.'

One letter that Pfeiffer wrote to Darwin, in 1871, not long after the publication of *Descent*, has survived; it shows that she was a keen reader of his work.

Mayfield, West Hill, | Putney, S.W.

Sir,

I have been reading your work on the "Descent of Man" with absorbing interest. Forgive me if as a stranger I offer a remark on that part of it which relates to the decoration of birds. I think I do not err in imagining that you yourself feel some diffidence in crediting these creatures with the high æsthetic instincts needful to account for ornamentation such as that found on the wings of the Argus pheasant, if the sense of beauty is assumed to be the sole worker towards this end. Could it not be that beauty, when of a nature thus recondite, has been only an incidental result, while the end towards which sexual selection has directly tended has been the perfecting of characters calculated simply to fascinate or allure? It seems to me that this is not a too nice distinction: that beauty does not necessarily fascinate, & that fascination does not always imply beauty; furthermore that powers far less advanced are needed for the subjection of the faculties to a *spell* than for any degree of deliberate appreciation. That the lower animals are preeminently liable to fascination in this restricted sense is shown in the paralysing effect of the eyes of snakes. Might not the plumage of the male Argus pheasant, with its balls trembling in their dusky sockets, exercise upon the female bird when cunningly exhibited before her, a sort of glamour akin to this? Fascination inviting in the one case to death, in the other to love & life, may be supposed to be painful or pleasurable according to its object.

I will not further intrude upon time which is so nobly occupied, but remain Sir | Yours with much esteem | Emily Pfeiffer

Down. | Beckenham | *Kent. S.E.*
April 26

Madam

I am much obliged for your kindness in writing to me. I think it w^d. have been an advantage if I had used the word fascination, but I intended to express some such idea when I join to charm admire &c, the word excite.—

Pfeiffer

B171

Mayfield, West Hill,
Putney, S.W.

Sir,

I have been reading your work on the "Descent of Man" with absorbing interest. Forgive me if as a stranger I offer a remark on that part of it which relates to the decoration of birds. I think I do not err in imagining that you yourself feel some diffidence in crediting these creatures with the high aesthetic instincts needful to account for ornamentation such as that found on the wings of the Argus pheasant,

Letter from Emily Pfeiffer. DAR 174: 40.
By permission of the Syndics of Cambridge University Library.

I fear that it w.^d be very rash to use the illustration of the snake, as very few naturalists believe in snakes having any such power; though I myself am inclined to be a believer.

Madam | Your obliged servant | Ch. Darwin

Darwin's niece Frances Julia Wedgwood, known as Snow, was one of the great intellectuals of Victorian England. She had little formal education, and much of her work had to be done early in the morning, before her domestic duties began; she cared for an invalid brother, and later for her elderly father (the Hensleigh Wedgwood who barely made it to his chaotic wedding in chapter 2). She taught herself Latin, Greek, French, and German, and wrote two novels in her mid-twenties. In 1861, she wrote a long article, 'The boundaries of science', on the theological significance of *Origin of species*. Darwin commented, 'I must tell you how much I admire your Article; though at the same time I must confess that I could not clearly follow you in some parts, which probably is in main part due to my not being at all accustomed to metaphysical trains of thought. I think that you understand my book perfectly, and that I find a very rare event with my critics' (*Correspondence* vol. 9, letter to F. J. Wedgwood, 11 July [1861]). According to her obituary in the *Spectator*, on 10 January 1914, Snow, like Cobbe, tried without success to get Darwin to read Kant. Her most radical belief, according to the obituary, was that the best part of truth was that which evaded our reason, and the best part of thought that which evaded our speech; like Kant, she was a powerful reasoner and an inexorable critic of reason. Darwin relied on her for help with logic, ethics, and Latin and Greek, although none of the Darwins seems to have sympathised with her more mystical side, or much enjoyed her writing. George Darwin said of her review of Samuel Butler's *The fair haven*, 'it was the most unintelligible mystical thing I ever read' (DAR 210.2: 31). Snow was a frequent visitor to Down, and her face-to-face discussions with Darwin are lost; this letter to Henrietta, however, carries some of the flavour of her conversation.

Tromers
April 1/1871

Is this the right address

My dear Harrot

I hope you will not get quite sick of the sight of my handwriting, but I want you to act as a funnel to your father for so much of my remarks as you may deem suitable. I never like to say all I have to say to him, for fear of tiring him. I feel I have a number of heterogeneous remarks to make some of them probably not worth making, but I sh^d like to tumble them all out of my mind. I gave him a little Abstract I had made of what I took for his views on Ethics, & I understood him to say he accepted them entirely so far as all except the self-regarding virtues went. Now I so want him to abolish that distinction between self-regarding & others-regarding virtues. (This is not a criticism of anything in his book only of this remark to me.) It is like the governess's division of plants into shrubs, bulbous plants & weeds. (the division is much better but I cant get hold of it exactly.) Of course in the police regulations of a nation we have to recognise some sort of distinction of that kind just as a gardener has to recognize the distinction between flowers & weeds, but there is surely no more science in one classification than in the other. I suppose temperance is a typical example of a self-regarding virtue, & yet if drunkenness did not put one out of the way of doing any good to one's fellow creatures & into the way of doing them a great deal of harm (I suppose no single fact is a larger source of crime) I cannot see what harm there w^d be in it. I do so thoroughly feel *nothing* is self-regarding in that sense. We radiate whatever we are.

I wanted to say something about Mr Morley's letter, which interested me very much—& yet I dont think there was much in it for Uncle Ch. The drift of it seemed to me, as far as I cd take it in without remembering the criticism from which the correspondence took its rise, a vindication of the intellectual character of what we mean by the sense of beauty, as an essential part of what he calls the Association philosophy. It was much as if in some chemical work you shd find some compound spoken of by the name which you wished to keep for one of it's constituent elements. I shd hardly think it was worth Uncle Ch's while to attend to a matter of nomenclature like that [*1 page or more missing*]

...

Yrs FJW

Emma wrote to Henrietta on 4 December 1873 about another meeting:

Eliz. & Snow dined ☉ [*yesterday*] & it is so m. better when we are a v. small party. F. [*father*] quite enjoyed his talk w. her— Over Max Müller & the nonsense he talks about language & reason. She has written down her confutation of him so clearly that one wonders how she can be so confused when she prints. I believe he will make use of it, but Snow thinks M. Müller so rubbishy she does not think him worth powder & shot— She has also given him a capital translation of Linneus acct of a sleeping plant, & F is quite pleased w. his own proficiency in Latin that now he has the translation he can quite make out the original.

Snow was acutely aware of her lack of education and opportunities, and wrote in the *Woman's Herald*, 23 May 1891 (cited in *ODNB*), 'My life ought to have been so much more than it has been.'

The Darwins were rather more in awe of another great female intellect of the age, George Eliot, at the time known familiarly as Mrs Lewes, although she and George Henry Lewes were not married, as the Darwins were aware. (Snow Wedgwood knew her and wrote about her in *Contemporary Review*.) Darwin enjoyed her novels; in 1873, he visited Mr and Mrs Lewes (George Lewes had become a correspondent of his, after writing a review of *Variation under domestication*), then begged an invitation for his daughter and her husband. Many people of the Darwins' class would have refused to visit Eliot, because of her relationship with Lewes.

16, Montague St | Portman Sq
March 30th.—

My dear M.^rs. Lewes.

I hope that you will forgive me for venturing to beg a great favour. Yours & Mr Lewes' kindness towards me when I have called is my sole & rather poor excuse. My eldest daughter & her husband, Mr Litchfield, have the strongest wish to be allowed to call on you some Sunday evening.— I think that it will be some recommendation of Mr Litchfield when I tell you that he has aided in every possible way during many years the Working Mens College; having first taught mathematiks & of late music & singing.— If you will grant your permission, the briefest line on a Post-Card will suffice. If I do not hear, I will understand that you have, as is too probable, already too many callers.

I beg leave to remain like so many other englishmen & english women | Yours very truly obliged | Charles Darwin

My wife complains that she has been very hardly treated, & that I ought to have asked permission for her to call on you with me when we next come to London; but I tell that I still have some shreds of modesty.—

The Priory, | *21 North Bank,* | *Regents Park.*
Mar. 31. 73

My dear Mr Darwin

We shall be very happy to see M.^r & M.^rs. Litchfield on any Sunday when it is convenient to them to come to us.

Our hours of reception are from ½ past two till six, & the earlier our friends can come to us, the more fully we are able to enjoy conversation with them.

Please do not disappoint us in the hope that you will come to us again, & bring M.^rs. Darwin with you, the next time you are in town.

Yours most sincerely | M E Lewes

Arabella Burton Buckley was a populariser of science. She worked as Charles Lyell's secretary from 1864 until his death in 1875, lectured on natural science, and revised Mary Somerville's *Connexion of the physical sciences* in 1877. (For Buckley's friendship with Alfred Russel Wallace, see

chapter 4.) She was the daughter of a clergyman, and married in her for-
ties. Her first book, *A short history of natural science*, was published in 1876;
Darwin sent this letter of praise.

Down Beckenham
Feb. 11th.

My dear Miss Buckley
　You must let me have the pleasure of saying that I have just finished reading with very great interest your new book. The idea seems
to me a capital one and as far as I can judge very well carried out.
There is much fascination in taking a bird's eye view of all the grand
leading steps in the progress of science. At first I regretted that you
had not kept each science more separate; but I daresay you found it
impossible.— I have hardly any criticisms, except that I think you
ought to have introduced Murchison as a great classifier of formations, second only to W. Smith. You have done full justice, and not
more than justice, to our dear old Master, Lyell.— Perhaps a little
more ought to have been said about Botany, and if you should ever
add this, you would find Sachs' History, lately published, very good
for your purpose.
　You have crowned Wallace and myself with much honour and
glory. I heartily congratulate you on having produced so novel and
interesting a work, and remain,
　My dear Miss Buckley | Your's very faithfully | Ch. Darwin

1 St Mary's Terrace | Paddington W.
Feb 12. 1876.

Dear M.ʳ Darwin,
　Thank you so much for your kind appreciative letter. It has given
me very great pleasure for I never expected that you would even read
such a simple little book as mine, much less read it through & take
the trouble to comment upon it. I have been thinking that I ought to
have said something of Murchison & now you suggest it I will work
it in to the next edition— M.ʳˢ Lyell [*Katherine Murray Lyell*] also had
already lamented to me the absence of botany in the 19.ᵗʰ century & I
am very glad to know of a book to which I can apply for information.
I tried to say what I thought was true about Natural Selection but it
was impossible to do it justice (even for beginners) in a few pages &
without details.
　Would you take the trouble on p 322 line 17 to substitute the words
"an incandescent gas" for "an ordinary gas flame" It is not true as it
stands but one of those stupid mistakes one discovers afterwards—
　I hope you are feeling well & able to do a fair share of work
　I often miss not hearing how you are & what you are doing, as I
did in past times—

Arabella Burton Buckley. From *Research: a monthly illustrated journal of science*, 1
February 1889.
The Bodleian Libraries, The University of Oxford, Per.1991 d.26, facing p. 130.

Please remember me very kindly to M⁏ Darwin & with many
many thanks for your kind letter

Believe me | Yours very sincerely | Arabella B Buckley

PS. Have you seen Monteiro's "Angola & the River Congo" I made
his acquaintance at the British Museum & found him a very intelligent observer. He says he has now had those copper-tailed birds in
confinement two or three years & the copper still appears on the tail
& is easily washed off—

In 1880, Buckley sent Darwin her book *Life and her children: glimpses of animal life from the amoeba to the insects*. In it, she explained evolution by natural
selection in terms of richness and opportunity: 'Since the whole world is
teeming with life, and countless numbers of seeds and eggs and young
beginnings of creatures are only waiting for the chance to fill any vacant
nook or corner, every living thing must learn to do its best, and to find the
place where it can succeed best and is least likely to be destroyed by others'
(p. 6).

Down Beckenham
Nov. 14ᵗʰ. 1880

My dear Miss Buckley

I am very much obliged to you for sending me your new book, the
appearance of which is most elegant. I have read the two first chapters and shall hereafter read more; but just at present I have a lot of
papers to read on account of work in hand.

I think that you have treated evolution with much dexterity and
truthfulness; and it will be a very savage heretic-hunter who will per-
secute you. I daresay that you will escape, and you will not be called
a dangerous woman.— Your plan seems to me an excellent one, and
who can tell how many naturalists may spring up from the seed sown
by you.— I heartily wish your book all success. At p. 4 I think you
ought to except utter deserts, for I believe they support nothing.—
I believe that you might make an equally interesting book for the
young about Plants.

Pray believe me, my dear Miss Buckley, your's sincerely Ch. Darwin

After Darwin's death, Buckley went on to publish *Winners in life's race* (about
vertebrates) in 1882, and *Moral teachings of science* in 1891. Like Snow Wedgwood and France Power Cobbe, she had an abiding interest in the development of moral qualities and mutual aid, seeing these even among plants,
who unconsciously co-operated with insects and punished cheats. To critics who complained that Darwinism displayed a world full of bitter competition and suffering, she retorted that the suffering and strife had always
existed; natural selection at least showed that it had a point, and further

that it was a consequence of universal law, expressing the will of God, not of caprice. Survival, in the animal kingdom, depended on ever-more conscious care for others, beginning with the care of parents for their young; and in human society, 'unfair advantage and hurtful actions toward the community create opposition which is a barrier to success' (Buckley 1891, p. 90).

11   Religion

Religion was an important realm of authority for women in the Victorian period, despite their exclusion from the official hierarchies of the Anglican church and most dissenting denominations. Though widely regarded as inferior to men in intellect, women were considered equal, or even superior, in moral conduct. This assumption rested in part on their allegedly greater capacity for sympathy, and it opened leadership roles for them in areas such as philanthropy, education, and religious authorship. Depending on their wealth and social status, women might be expected to provide for the relief and improvement of the poor in the parish, or to teach in the village Sunday school. The home was also a place of religious devotion. As wives and mothers, women played a central role in the care of the household, in family prayer and the reading of Scripture, and in the moral upbringing of their children.

The Darwins and Wedgwoods were nominally Anglican. Formal adherence to Church of England doctrine was especially important for men, being an essential condition for admittance to many elite institutions, such as Oxford and Cambridge, which were important gateways to the most prestigious professions. In personal belief and practice, however, the families were Unitarian. Greater emphasis was placed on inner feeling and moral conduct than on doctrine, though belief in the afterlife remained important. The importance of prayer, Bible reading, and religious feeling are evident in letters from Darwin's sister Caroline, and his cousin and future wife, Emma Wedgwood. Tensions arising from Darwin's heterodoxy and immersion in scientific pursuits were addressed in correspondence between the couple after their engagement in November 1838. Other family letters discuss personal devotion in relation to ritual practices such as churchgoing and catechism, and provide a glimpse of Emma's moral role as educator of her children. The Darwins were active supporters of the church in Down, and on friendly terms with the local Anglican clergy, especially John Brodie Innes, who was the perpetual curate in the village from 1846 to 1868. Several years after his departure, a conflict of authority arose between the Darwins and the new vicar. Letters from Emma between 1873 and 1875 discuss this ongoing dispute, and convey her general dissatisfaction with parish affairs, and a shift of allegiance toward Protestant dissent.

The implications of Darwin's evolutionary theory for religion are usually examined with reference to public debates from which women were largely excluded. In letters, however, women engaged with Darwin's work directly, raising questions that confronted them in their roles as mothers, educators, and religious writers. Darwin's replies to such letters were reassuring but brief, and he was extremely reticent on religious matters in print. After his death, controversy arose over an extended discussion of Christianity that he had included in a private memoir, and that some members of the family wished to publish as part of an extended biography. Thus tensions that had remained submerged during Darwin's lifetime resurfaced in negotiations over his public image and legacy. Emma sought to preserve for posterity the moral character of the man she had loved, cared for, and admired.

The importance of religious belief and practice in the Darwin family is indicated in a letter from Darwin's sister Caroline, who wrote from Shrewsbury on 22 March 1826, describing a party of old friends, one of whom, William Vaughan, was a vicar in Shropshire.

My dear Charles

…

On Monday we had a party of Erasmus's friends—Hildyard, Wakefield, Wingfield & Vaughan, I liked them all except the latter— whose manners I thought as bad as many of his opinions—he evidently rather laught at the suffrings of the poor slaves & admires some book which John Bull [*a popular Tory weekly*] quotes which is I suppose taking the slave holder's part— Moreover, he seemed ashamed of doing his duty as a clergyman, & said *packs of cards* he should be more likely to purchase than bibles & prayer books, & rather implied all *very* religious people were hypocrits, &c. I sat between him & Hildyard who I like very much & who when he saw I would not answer Vaughan when he laughed on religious subjects immediately joined in the conversation & so pleasantly that he quite confirmed my liking to him.—

dear Charles I hope you read the bible & not only because you think it wrong not to read it, but with the wish of learning there what is necessary to feel & do to go to heaven after you die. I am sure I gain more by praying over a few verses than by reading simply— many chapters— I suppose you do not feel prepared yet to take the sacrament—

In his reply on 8 April, the young medical student at Edinburgh shows his respect for his elder sister, and alludes to her role in his upbringing. The moral influence of his sisters is a point he would repeat in his later autobiographical reflections.

April 8th

My dear Caroline

I dare say I shall not be able to finish this letter, but I cannot help writing to thank you for your very nice and kind letter. It makes me feel how very ungrateful I have been to you for all the kindness and trouble you took for me when I was a child. Indeed I often cannot help wondering at my own blind Ungratefulness. I have tried to follow your advice about the Bible, what part of the Bible do you like best? I like the Gospels. Do you know which of them is generally reckoned the best? Do write to me again soon, for you do not know how I like receiving such letters as yours.

Caroline wrote back on 11 April 1826, recommending the gospel of John and the epistles of John and James, adding: 'I often regret myself that when I was younger & fuller of pursuits & high spirits I was not more religious—but it is very difficult to be so habitually' (DAR 204: 24).

Darwin's divergence from orthodox Christianity is evident in some of his notebooks from the late 1830s. Upon his engagement to Emma Wedgwood in November 1838, he was advised by his father to keep his religious views private. He did not heed this advice, however; and the couple worked to reconcile their differences of belief in the months just before and after their marriage. Emma wrote to her fiancé from her father's house in Shrewsbury on 23 January 1839, in anticipation of their coming marriage.

> Maer
> Wednesday

My dear Charles

… You need not fear my own dear Charles that I shall not be quite as happy as you are & I shall always look upon the event of the 29th as a most happy one on my part though perhaps not so great or so good as you do. There is only one subject in the world that ever gives me a moments uneasiness & I believe I think about that very little when I am with you & I do hope that though our opinions may not agree upon all points of religion we may sympathize a good deal in our *feelings* on the subject. I believe my chief danger will be that I shall lead so happy comfortable & amusing a life that I shall be careless & good for nothing & think of nothing serious in this world or the next. However I won't be solemn either.

A longer letter followed, written some time in February. Darwin's side of the correspondence has not survived, only a note added to the end of this letter: 'when I am dead, know that many times, I have kissed & cryed over this.'

> The state of mind that I wish to preserve with respect to you, is to feel that while you are acting conscientiously & sincerely wishing, & trying to learn the truth, you cannot be wrong; but there are some reasons that force themselves upon me & prevent my being always able to give myself this comfort. I dare say you have often thought

of them before, but I will write down what has been in my head, knowing that my own dearest will indulge me. Your mind & time are full of the most interesting subjects & thoughts of the most absorbing kind, viz following up yr own discoveries—but which make it very difficult for you to avoid casting out as interruptions other sorts of thoughts which have no relation to what you are pursuing or to to be able to give your whole attention to both sides of the question.

There is another reason which would have a great effect on a woman, but I don't know whether it w^d so much on a man— I mean E. [*Erasmus*] whose understanding you have such a very high opinion of & whom you have so much affection for, having gone before you— is it not likely to have made it easier to you & to have taken off some of that dread & fear which the feeling of doubting first gives & which I do not think an unreasonable or superstitious feeling. It seems to me also that the line of your pursuits may have led you to view chiefly the difficulties on one side, & that you have not had time to consider & study the chain of difficulties on the other, but I believe you do not consider your opinion as formed. May not the habit in scientific pursuits of believing nothing till it is proved, influence your mind too much in other things which cannot be proved in the same way, & which if true are likely to be above our comprehension. I should say also that there is a danger in giving up revelation which does not exist on the other side, that is the fear of ingratitude in casting off what has been done for your benefit as well as for that of all the world & which ought to make you still more careful, perhaps even fearful lest you should not have taken all the pains you could to judge truly. I do not know whether this is arguing as if one side were true & the other false, which I meant to avoid, but I think not. I do not quite agree with you in what you once said—that luckily there were no doubts as to how one ought to act. I think prayer is an instance to the contrary, in one case it is a positive duty & perhaps not in the other. But I dare say you meant in actions which concern others & then I agree with you almost if not quite. I do not wish for any answer to all this—it is a satisfaction to me to write it & when I talk to you about it I cannot say exactly what I wish to say, & I know you will have patience, with your own dear wife. Don't think that it is not my affair & that it does not much signify to me. Every thing that concerns you concerns me & I should be most unhappy if I thought we did not belong to each other forever

References to religion in letters between the couple are very infrequent thereafter. After one of Darwin's long periods of illness in June 1861, Emma wrote:

I cannot tell you the compassion I have felt for all your sufferings for these weeks past that you have had so many drawbacks. Nor the gratitude I have felt for the cheerful & affectionate looks you have given me when I know you have been miserably uncomfortable.

My heart has often been too full to speak or take any notice   I am sure you know I love you well enough to believe that I mind your sufferings nearly as much as I should my own & I find the only relief to my own mind is to take it as from God's hand, & to try to believe that all suffering & illness is meant to help us to exalt our minds & to look forward with hope to a future state. When I see your patience, deep compassion for others self command & above all gratitude for the smallest thing done to help you I cannot help longing that these precious feelings should be offered to Heaven for the sake of your daily happiness. But I find it difficult enough in my own case. I often think of the words "Thou shalt keep him in perfect peace whose mind is stayed on thee". [*Isaiah 26:3*] It is feeling & not reasoning that drives one to prayer. I feel presumptuous in writing thus to you.

I feel in my inmost heart your admirable qualities & feelings & all I would hope is that you might direct them upwards, as well as to one who values them above every thing in the world. I shall keep this by me till I feel cheerful & comfortable again about you but it has passed through my mind often lately so I thought I would write it partly to relieve my own mind.

Again, there is no reply from Darwin, only this annotation at the end of the letter, 'God Bless you. C. D.'

The Darwin family attended the local Anglican church of St Mary's each Sunday, as there was no Unitarian chapel in Down. Though Darwin stopped going to church around 1850, Emma continued to take the children, and all of them were baptised and confirmed in the Church of England. On 6 May 1859, she wrote to her eldest son, William, who had settled in Southampton as a banker, about the importance of religious duty and devotion in relation to outward forms of Christianity, such as churchgoing.

It was very nice of you going to Church on Sunday in the midst of your bustle, because you saw I wished it, but I should be very sorry if you got to consider going to Church on Sunday as only a decent form (which may be put aside for a small reason), & not a real duty. I think the daily Chapel is very injurious in almost inevitably making you so weary of the service & taking away the solemnity of the prayer &c. I hope you will try to consider the Sunday service as a different & much more real duty.

Goodbye my dear old man.

On 26 September 1866, Emma wrote to Henrietta about her youngest daughter, Elizabeth, aged 19, who was about to be confirmed, and who had reservations about reciting the catechism.

> There is to be a confirmation in Nov. & I want you to try to remember the sort of exam. Mr S. [*Thomas Sellwood Stephens, curate of Down*] gave you, & especially how much Cat? Lizzy says she sh^d feel hypocritical to have any thing to do with the Cat. & that as she does not believe in the Trinity or in Baptism she does not feel much heart for it. The per contra is that I think it would give her a zest in searching in the Bible (I remember thinking it was good for you) & though the doctrine w^d most likely slip over her mind without any impression I think the ceremony itself is impressive & simple. Otherwise I have no doubt Mr S. w^d give her the Sacrament without enquiry. Please write soon. She half wishes it herself if she c^d get rid of the Cat.

Three years later, in the winter of 1869–70, Elizabeth wrote to her brother Horace about the new vicar of Down, Henry Powell, and his plan to open an evening reading-room for working men in the village as an alternative to the public house.

> Monday.
>
> Dear Horace.
>
> Lenny was here yesterday, but George not as he was at Cambridge. It was a horrid cold day, and went a walk through snowy fields. Pauline Kilian is gone now I was very sorry when she went for she was very pleasant and I liked to have somebody to take walks with. Mr Powell has been calling here he is a nice little man, I think we are very lucky to have such a nice clergyman he seems so zealous about the school and everything. He is going to set up a reading room in the village, which I am very glad of. It is to be in the school, from six till nine every day, there are to be penny newspapers, and Coffee which they are to buy, they are to be allowed smoking which is the great thing to make them comfortable. I hope it may succeed, Mr Powell seems rather doutful, he says that so many of the men cannot read.

This reading-room would become a pet project of the Darwins, and a point of contention after Powell's departure. On 22 November 1873, Emma wrote to the new vicar, George Ffinden, in hopes of gaining his support for continuing the venture. The Darwin and Lubbock families had long supported the church through charitable activities, and Darwin had taken on additional financial and administrative responsibilities in the late 1860s, when several of the local vicars had failed to fulfil their duties. Emma regarded the provision of a reading-room for working men as a matter of religious and social duty.

Elizabeth Darwin. DAR 225: 74
By permission of the Syndics of Cambridge University Library.

> *Down, | Beckenham, Kent.*
> Saturday

Dear Mr Ffinden

Sir John & Lady Lubbock as well as M[r] Darwin & myself are anxious to establish the Reading Room for the winter months as we did formerly, & I shall be much obliged to you to inform me whether I ought to apply to the School Board for permission to use the School room for that purpose.

My application would also have much greater weight if you were inclined to join in the request.

We should of course have to pay a woman to put the schoolroom in complete order every morning before the school assembles.

With kind regards to Mrs Ffinden

very truly yours | Emma Darwin

Ffinden opposed the measure, however, and took issue with the Darwins for their involvement in parish affairs, which he considered as an affront to his own authority. Emma wrote to Horace on 29 November 1873 of the trying affair.

> We are now fighting Mr Ff– & Mr Allen [*the churchwarden*] tooth & nail about having the school room for a Reading room. There first objection was the smell of smoke which they said M[rs] Laslett & Mr Pearson [*local schoolteachers*] found so annoying. This turns out to be an invention as we have enquired of both, so their new dodge is that we shall lose the capitation grant if the school is used for any thing. So we have applied to the fountain head (wherever that is— Richard knows) but we are convinced it is another invention. I don't know why Mr Allen is so zealous as he is going away; perhaps because Frank will not take any of his furniture (except what he is obliged to do & one or 2 articles)— [*Francis was planning to move into Allen's old house in Down.*]
>
> How it does rain.
>
> Good bye my dear old man E.D—

Relations with Ffinden did not improve, and Emma began to transfer her allegiance to the dissenting chapel, which she regarded as doing better service to the village. She wrote these letters in 1874 and 1875 to a former vicar of Down and family friend, John Brodie Innes, about the growing support for evangelical Protestantism, including a local temperance society, whose leaders had taken responsibility for the reading-room.

> *Down, | Beckenham, Kent. | Railway Station | Orpington. S.E.R.*
> Oct 12—

My dear Mr Innes

It is some time since I have sent you any parish news; & as we hear a rumour of an important change, I will give you the first information of it; though I have not much hopes that it will prove true.

It is said that on the death of Mr Ffinden's uncle Mr Sketchley, vicar of Deptford, which has just taken place, his son is to come here & Mr Ffinden to take the Deptford living.

This would certainly be a great blessing to this place, as Mr Ffinden has no influence here & has excited general dislike. The chapel is so crowded that it has been enlarged. I do not mention this as an evil from my point of view, but only as a proof of M^r Ff's unpopularity. You will not think me an impartial person perhaps as he cuts every member of our family when we meet; but as I said before the scheme of exchange sounds most improbable.

Down Beckenham

Dec 24—

Dear Mr Innes

...

The great event last week was the opening of a Reading Room, when Mr Nash gave a good supper to whoever chose to come & I was not surprized to hear that he had 90 guests. They have hired George Wood's old house for the purpose & begin the world with 45 members. Of course they will not nearly pay their way; which one would have preferred. We have also a band of Hope under Mrs Nash's superintendence which is of course prosperous at present, while the children are young & have no temptation; but I have some hopes that the effect may remain with some, especially of the girls, after they are grown up. Both these undertakings are thorns in Mr Ffinden's side & he has not been content with holding aloof from them; but has used all his influence to prevent their succeeding.

Darwin published very little about the implications of his evolutionary theory for religion. Correspondence was thus an important medium through which readers of his work tried to engage him on religious subjects. One of the most searching letters Darwin received was from Mary Boole. At the time of writing in 1866, she was a widow, supporting her five daughters as a librarian at Queen's College, London, the first women's college in England. She also gave Sunday evening talks at the college, in which she discussed the relationship of different forms of knowledge. She would later write a book, *The message of psychic science to mothers and nurses*, that addressed some of the challenges posed to religious belief by Darwin's theory. In his reply, Darwin suggested that science and religion rested on different foundations or forms of evidence, and alluded to the possible bearing of the theory of natural selection on the problem of evil.

Private

Dear Sir

Will you excuse my venturing to ask you a question to which no one's answer but your own would be quite satisfactory to me.

Do you consider the holding of your Theory of Natural Selection, in its fullest & most unreserved sense, to be inconsistent,—I do not say with any particular scheme of Theological doctrine,—but with the following belief, viz:

That knowledge is given to man by the direct Inspiration of the Spirit of God.

That God is a personal and Infinitely good Being.

That the effect of the action of the Spirit of God on the brain of man is *especially* a moral effect.

And that each individual man has, within certain limits, a power of choice as to how far he will yield to his hereditary animal impulses, and how far he will rather follow the guidance of the Spirit Who is educating him into a power of resisting those impulses in obedience to moral motives.

The reason why I ask you is this. My own impression has always been,—not only that your theory was quite *compatible* with the faith to which I have just tried to give expression,—but that your books afforded me a clue which would guide me in applying that faith to the solution of certain complicated psychological problems which it was of practical importance to me, as a mother, to solve. I felt that you had supplied one of the missing links,—not to say *the* missing link,—between the facts of Science & the promises of religion. Every year's experience tends to deepen in me that impression.

But I have lately read remarks, on the probable bearing of your theory on religious & moral questions, which have perplexed & pained me sorely. I know that the persons who make such remarks must be cleverer & wiser than myself. I cannot feel sure that they are mistaken unless you will tell me so. And I think,—I cannot know for certain, but I *think*,—that, if I were an author, I would rather that the humblest student of my works should apply to me directly in a difficulty than that she should puzzle too long over adverse & probably mistaken or thoughtless criticisms.

At the same time I feel that you have a perfect right to refuse to answer such questions as I have asked you. Science must take her path & Theology hers, and they will meet when & where & how God pleases, & you are in no sense responsible for it, if the meeting-point should be still very far off. If I receive no answer to this letter, I shall infer nothing from your silence except that you felt I had no right to make such inquiries of a stranger.

I remain | Dear Sir | Yours truly | Mary Boole

43 Harley Street | London W.

Dec. 13th. 1866

Down. Bromley. Kent.
Dec! 14. 1866.

Dear Madam.

It would have gratified me much if I could have sent satisfactory answers to y! questions, or indeed answers of any kind. But I cannot see how the belief that all organic beings including man have been genetically derived from some simple being, instead of having been separately created bears on your difficulties.— These as it seems to me, can be answered only by widely different evidence from Science, or by the so called "inner consciousness". My opinion is not worth more than that of any other man who has thought on such subjects, & it would be folly in me to give it; I may however remark that it has always appeared to me more satisfactory to look at the immense amount of pain & suffering in this world, as the inevitable result of the natural sequence of events, i.e. general laws, rather than from the direct intervention of God though I am aware this is not logical with reference to an omniscient Deity— Your last question seems to resolve itself into the problem of Free Will & Necessity which has been found by most persons insoluble.

I sincerely wish that this note had not been as utterly valueless as it is; I would have sent full answers, though I have little time or strength to spare, had it been in my power.

I have the honor to remain dear Madam. | Yours very faithfully | Charles Darwin.

P.S. I am grieved that my views should incidentally have caused trouble to your mind but I thank you for your Judgment & honour you for it, that theology & science should each run its own course & that in the present case I am not responsible if their meeting point should still be far off.

43 Harley S!
Dec! 17ᵗʰ.

Dear Sir,

Thank you sincerely for your kind letter. You have told me all I wanted to know from you. The criticisms to which I referred were such as seemed to take for granted that all such speculations as yours,—in fact, as it seemed to me, *all* independent un-theological speculations on Creation as we find it,—must be incompatible with any belief in a moral government of the world. I have always taken the liberty of telling the people who brought such criticisms under my notice, that, in my opinion, the authors of them were simply talking about what they had never examined into. But still, when one is studying alone, & so ignorant too as I am, one gets frightened, & loses faith in one's own principles. And I thought, for my own satisfaction, I should like to have *your* assurance that moral & religious faith are things quite independent of theories about the *process* of Creation. You have given me that assurance and again I thank you.

　With sincere wishes for improvement in your health

　I remain | dear Sir | Yours truly | Mary Boole

Other correspondents shared some of the inner struggles that arose from reading Darwin's work. An Austrian, Mary Jung, about whom nothing else is known, wrote of her effort to reconcile evolutionary theory with Christianity. Darwin tried to reassure her that theology would eventually be brought into harmony with scientific truth.

　I dont know a man, whoes opinion produced such a great revolution in this branche of science, a theory, which found such a general embracing, inspite the greatest efforts of refutation from other parts. I am partly submitted to your opinion and I remain doubting between your theory and the ecclasiastical dogma. When my reason agrees with your opinion, my heart stands to the latter and so I am in a continnual conflict with myself. I beg to excuse my speaking to you so freely, and I hope you will therefor not be unfavourable to me. You would render *very very* happy by affording the request of honouring me with a single line.

　Yours | most thankfully and humbly | Mary Jung

Villa Jung | Salzburg, 7. 1. 79.

Darwin replied:

　Permit me to advise you to try not to be troubled about the differences between ecclesiastics & scientific men. Search for the truth, & then your conscience will be at ease. In the course of time ecclesiastics have always managed to make their conclusions somehow to harmonise with ascertained truths, which they at first vehemently & ignorantly opposed

Darwin had speculated on the natural history of ethics and religion since the beginning of his work on species in the late 1830s. When he finally came to write about human evolution in *Descent of man* (1871), he briefly discussed the origin of certain religious feelings and beliefs, and devoted a chapter to the development of the moral sense through natural selection. Darwin discussed some of this controversial work with members of the family, including his niece Frances Julia Wedgwood (Snow). He also gave the draft of his chapter on morals to Henrietta for her critical commentary (see chapter 9). A letter from Emma to Henrietta on 8 February 1870 reveals that religion was still a sensitive and potentially divisive issue within the family.

F. [*father*] is hard at work on the moral question of man & had talks w. Snow about defining religious feeling, in w. she only admitted love & reverence & left out fear, but owned she was mistaken after all. F. is deeply interested in the question & I wish it was over as it absorbs him too much & he had to lie by one day.

And on 21 February:

Mr Dulin is copying F's Chapter & as soon as he has looked it over again I believe he means to send it to you at Cannes. I think it will be v. interesting but that I shall dislike it v. m. as again putting God further off.

Shortly after *Descent* was published, in 1871, Emma expressed reservations about Darwin's moral theory to Frances Power Cobbe, a religious writer and family acquaintance. Cobbe had published a review of *Descent* in the *Theological Review*; in it she criticised Darwin's too sanguine view of human nature. (The Darwins had expected her to write for her usual paper, the *Echo*.) Darwin argued that angry or selfish feelings, though strong, were of comparatively short duration, and that these were naturally overcome by benevolent instincts of love and sympathy for others, which had evolved in social animals and persisted in humans, forming the basis of moral conduct.

6 Queen Anne St
Saturday

My dear Miss Cobbe

Many thanks for your taking the trouble to write. Mr Murray seemed to think that he had made an unreasonable fuss in the matter. We want to see the Echo, tho' not so much as if you had written it. The papers we have seen are quite mild & civil.

M⁣ʳ Darwin says that he knows so well how much you & many others will disapprove of the moral sense part that he will not be surprised at any degree of vigour in your attack.

He does not know to what you refer when you say that he does not distinguish between regret & repentance or remorse, he remembers reflecting on the wide & obvious distinction. It appears to him that as long as hatred is felt against any one, the social instincts are over-mastered & there is no room for repentance. But no doubt he will understand what you mean when he reads your article.

If you should come to Mʳ Martineau's tomorrow do look in at No 6 Q. Anne St where we are—

Speaking in my private capacity I quite agree with *you*. I think the course of all modern thought is "desolating" as removing God further off. But I do not know whether his views on the moral sense would *exclude* Spiritual influence though not included in his theory—

So you see I am a traitor in the camp.

With very kind regards to Miss Lloyd. | Yours very truly | Emma
Darwin

As his fame grew, Darwin was increasingly approached by individuals who
sought to attach his name and reputation to a social or political cause. This
was particularly true of movements of secular religion or free thought. In
1871, Darwin corresponded with the American journalist Francis Abbot,
who wished to quote several supportive remarks of Darwin's in his newspaper the *Index*, an organ for 'free religion'. Darwin was reluctant to have his
private statements quoted in public, and asked Henrietta for advice in an
undated letter that must have been written on 13 November 1871. She signs
her stern reply 'Rhadamantha', the name given her by Thomas Huxley,
based on the character from Greek mythology and denoting an inflexible
judge and severe master.

Monday

Dear Father

Thanks for Abbott letter. I'm sorry to say I don't much like its tone.
I call it flabby & it lowers my opinion of him. His sending the 50 dol-
lars too shows a great want of judgement.— However this is beside
the point— what I meant to beg you to consider very seriously is
whether it isn't a great pity that you should lend your name to *any religious movement* whatever. If you write any sentence to be printed it will
mean so m. more than that you have read through the Truths for the
Times once or twice & were much struck with them. I consider that is
in your private capacity just as Jones or Smith might be struck—but
printing it as the Author of the Origin it *ought* to & *will* mean very
much more than I think it has a right to mean— I don't want you
to commit yourself to any definite opinions in the religious scientific
questions. You have not time or strength to go in for it thoroughly. &
then speaking of it damages you as a scientific man & does not do a
compensating good to free thought— I feel so profound a conviction
that to let people draw their own inferences if they feel that you have
no party spirit will be such an infinitely stronger lever. I have felt
so thankful you were not as Tyndall— [*The physicist John Tyndall had
questioned the power of prayer and the existence of miracles.*] You ought to be
as careful of your fair fame as Cæsars wife or whomever it was. …

Forgive this cool lecture dear Father I'm too stupid to make it not
read so bald & dictatorial— I only mean it is my deliberate opinion—
Ever your faithful | Rhadamantha

The question of making Darwin's private views on religion public arose
again after his death. In 1876, Darwin had written a memoir, 'Recollections of the development of my mind & character', intending it for his

children and grandchildren. It contained a lengthy discussion of his religious beliefs, as they had evolved from orthodox Christianity, to theism, to agnosticism, as well as several critical remarks about Christian doctrine. Francis wished to include the memoir in his edition of his father's *Life and letters*, and a controversy arose within the family about whether to publish the section on religion as it was written, with some passages deleted, or to exclude it entirely. Francis worried that others, such as Ernst Haeckel and Edward Aveling, had already published on the religious tendencies of Darwinism, and he wanted to present his father's own words on the subject. Henrietta argued along the lines she had taken in her 1871 letter to her father, that his personal beliefs on religion were not of the same order as his scientific work. They were not the product of the same rigorous investigation and attention. They did no credit to him, and they opened avenues of misinterpretation and appropriation by others.

Henrietta voiced her opinion to her mother, who had her own reservations about publishing statements that she believed were ill-considered, and that failed to present Darwin's beliefs adequately. Emma expressed her misgivings in a series of letters to William in January 1885. William was the legal executor of the manuscript, and so had the final say as to whether the material would be published. In the end, a compromise was struck, with substantial material removed. It appears that Emma played the deciding role. Her sons all agreed that nothing should be published that would be painful to her, but this was clearly not the only factor that governed her excisions. She much preferred his letters to various correspondents, such as the Dutch student Nicolaas Doedes, who had written at length for Darwin's views on religion: 'I am aware that if we admit a first cause, the mind still craves to know whence it came and how it arose,' Darwin replied on 2 April 1873, 'The safest conclusion seems to be that the whole subject is beyond the scope of man's intellect; but man can do his duty.'

The Grove. | Huntingdon Road. | Cambridge.
Tuesday Jan 20 | 1885

My dear William

I feel much perplexed about the Auto. I enclose what Hen. says in the matter, which has not shaken Franks opinion at all— I think you & he are somewhat biassed by Haeckels insolence on the subject, & I think you shd consider what is best without any reference to that—

My present view is that Frank should tell in his own words the result that your father arrived at, but not give the whole process of thought, perhaps quoting several passages, & saying that he had written more fully on the subject— The first opening sentence about the God of the Old Testament appearing as (I quite agree w. him that it does so) a revengeful tyrant, would raise a storm of indignation especially as he omits the higher & more sublime views of God in Isaiah & some of the Psalms— & that sort of passage makes me feel that his view was narrow & wanted study—

There is also a passage about the supposed intuitive idea of God, which I think inconclusive & narrow, & not worthy of his mind— …

My dear William

… I am no nearer a settlement but so far; if it were not for the letters published & the conversation w. Aveling, I should prefer his opinions being only drawn from his works; but that being out of the question I am in favour of either publishing the Religious part with the exception of the beginning, or *what* I should prefer, Frank giving a short account of his opinions in his own words … without however approving of it— …

The Grove. | Huntingdon Road. | Cambridge.
Sunday—

My dear William

I will lock up the letters & carefully keep them— I agree with you that they are admirable, & whether the religious part of the Auto. is published or not I think they should be published. His state of uncertainty as to any positive believe seems to me to be better shewn than in the longer Auto—

I will in a few days send you my copy of that part, in which I have marked what strikes me as inconclusive or narrow—& I shall like to know your impression of the whole, as to whether it is altogether worthy of the powers of his mind in the argumentative part— Any how I should wish that his conclusions were made public— …

The Grove. | Huntingdon Road. | Cambridge.
Monday—

My dear William

I had a long talk w. Leo which has helped to settle my mind, & I quite agree to the publication of the Fragment with the exception of one passage—p. 42—beginning "But I had gradually become — — & ending "This appeared to me utterly incredible"

For myself I quite agree with these opinions but it would give such a violent revulsion & shock to most believers that I should feel it wrong without a very strong motive to do so—

Leonard wishes to cancel the whole paragraph, as far as "but as they influenced me", & I should prefer it too, but not to the same extent as the first part— The objections to Revelation in both parts are so obvious & well known, (as your father says) that I should like to omit them—

I should also like to omit (& Henrietta feels this strongly too) the short passage just after, ending "& have never doubted for a single second that my conclusion **was correct**". This sounds presumptuous

& I don't think it was what your father meant to express; but that he never for a single second altered his opinion on this part of the subject (viz the evidences of Christianity)— To say that ones conclusion is *correct* on such an immense subject as religion does strike me as dogmatic.

As we go against Hen's opinion in the whole matter I sh^d be glad if you c^d agree to this. I agree with her in thinking as far as I can judge that the whole is not what the world will think worthy of the highest powers of his mind—but Leonard has put before me his opinion of the great simplicity of the whole (I may say here that Leonard does not feel the least objection so omitting any part of this or of the whole Auto–B–, even without making any mark to shew that some part is omitted). It would require a few words by Frank such as "After detailing his disbelief in the Old Testament & the steps, he continues "I gradually came to disbelieve in Xianity as a divine revelation". I have made a mess of it but you know the sort of thing I mean—

About the letters which I return I admire most the one to Doedes, ending "but man can do his duty"— Leonard thinks that a disquisition on the being of God or immortality does not excite nearly such bitter feelings as a definite attack on the old or New Testament—

The passage to which Leo has put "stet", about the morality of the New Testament has given us many doubts— I sh^d be sorry to omit it on one account & yet I cannot understand it—

The alternative in my mind to omitting the 1st passage w^d be to omit the whole disquisition, or what I think you w^d prefer, the whole Auto. This w^d no doubt lessen the interest of the Life to a great degree. Frank w^d have to supplement the loss by an account of his early life written by himself, & the letters would tell his progress of feeling a great deal— Then the Auto– w^d be kept by you.

I think as all the brothers & sisters have been consulted George should not be omitted & I believe he will soon be at home— I shall now be able to cast the subject off my mind. …

In the end, Emma's concerns were heeded. The lengthy section on religion was removed in its entirety from the autobiography as it was published in Francis's *Life and letters*, with portions appearing in a separate chapter of the book on religion, together with a selection of letters that Emma thought better reflected Darwin's beliefs and feelings. The full version of Darwin's autobiography would eventually be published by his granddaughter, Nora Barlow, in 1958. (See also, more recently, 'Recollections', edited by James A. Secord.)

12   Travellers

Darwin's female friends, relations, and correspondents didn't tend to mount scientific expeditions of the sort that led men to write to Darwin for advice; nevertheless, in the quest for health, accompanying husbands to foreign postings, or decamping with their families to new homes abroad, they did travel, and sent back their impressions.

The female traveller who was closest to home was Henrietta. After suffering a great deal of illness in her youth, Henrietta recovered enough to take several trips to Europe. Emma worried, 'I only hope you may not turn into a regular travelling old maid living abroad' (DAR 219.9 73). She wrote long letters home ('I find it decidedly trying to read yr descriptions', wrote her forthright friend, Elinor Bonham Carter). She always travelled in company with a friend or a relation, staying at guest-houses patronised by other English people, and at a pinch a brother or servant could be dispatched to Paris or Calais to bring her home if the party broke up leaving her unchaparoned. Nevertheless, by 1870 she declared to Emma, 'an idiot c^d travel alone in France' (DAR 245: 40). In 1866, she was travelling in France with Elinor, who had been unwell and wanted to go abroad. In this letter of 4 June, she describes a train wreck near Arles to Emma.

> E. got better after Marseilles—19 o'clock where we had some hasty food—we managed to tip a porter & induced him to keep our carriage free so we made up our beds told the guard to wake us at Avignon & settled off to doze away the 3 hours—2 of them passed away in semi sleep & I believe I was just going right off when I started up feeling that we were off the line—tearing along & shaking us a good deal though not enough to knock us off our seats—Elinor thought of the bank at the side—I where to sit where I shd have the least chance of death—but bfore we had time for much thought, the train was stopped—450 steps a little soldier told Elinor it was. I shd hardly have thought it was so much— Happily for us the engine diverged to the right— ... we were in the 1st. carriage & should have all been killed or badly hurt—as it was, there was there was no harm done to any one person in the train— You may imagine how thankful we were to open our door & get on to firm ground uninjured! The cause of the accident was the "essieu" of the wheel breaking—i.e. the axle I imagine. We declined getting in to our carriage again & took out our cloaks & bags to sit on the

ground. Luckily it was very warm & a bright moon which was an immense comfort—also we were only 12 kilometres—about 8 miles from Arles where they sent at once for help—& navvies to begin getting the line in order— When we heard another train coming, as we did pretty soon, we determined to go down the embankment to get out of the way if there shd be any fresh scrimmage— The burnt child dreads cold water—for of course they did really stop it in time—so we knew by the difficulty we had in coming up again that the bank was pretty steep— I got so tired I lay down flat on the stones by about one oclock—our chief sufferings were from musquitoes— I had a veil, but got 23 bites on my hands and wrists— We then tried to get into our carriage again but the navvies had come & everybody told us we must wait a "petit moment" wh. in the language of that night meant about an hour— After we had sat about 2 hours a friend said he saw the "secours" & all the men rushed to look out—but he deceived us & another hour elapsed before we did thankfully bundle into a train & set off to Arles. Such a long 3 hours as it seemed! We had been more than 12 hours on the go & I was so tired, & it needs not to say E. was— Happily had neuralgia kept off till morning & then she had a bad enough attack to make up— I had this before me as a cheerful vision to keep up my spirits as I know she so often pays for things afterwards— At last we got to Arles & before we cd get into the waiting room every arm chair was full—& the atmosphere so awful we went outside on to some building stones & there sat dismally watching the dying moon on one side & the first streaks of light on the other— Oh! such a set as we all looked—blear eyed green & grimy & silent misery on every one's face— I got the windows in the waiting room open & we went there & sat on the hard wooden benches & envied some few of the men who snored around us in uncomfortable attitudes— At last after hours of waiting expecting every ¼ hour to be taken on to Avignon with our luggage we collared a chef & got the news that it was no use waiting for our luggage & that it wd be quite safe & we shd get it in time so at 6 o'cl. we got into a 'bus & went up to the nasty Arles inn—where we got some nasty breakfast & laid ourselves down on some nastier beds to wait for the midday train. I went to sleep—poor E. did not. All things have an end & at last we did get here—very late of course—having to wait for 5 luggage trains of interminable length to get on sidings & then having to travel with some children who eat sausage pie! little pigs—how I hated 'em! But we did get all our luggage which was a g! relief. I shd have thought at least it wd have gone on to Lyons— Then our nice clean room & water & soap & comf appetising beds made us feel quite diff. creatures. I have such an odd feeling about it all— As if it had happened all before— I seem as if I had had the precise feeling before of the carriage going off the line— It seems to me as if the other time was more real than this. I can't tell you want an odd feeling it is. I suppose you must have been hurt to make you feel nervous—but I don't feel as if this would make me the least bit more frightened which is a good thing—as being nervous in railways wd be very inconven! On the cannon ball principle I think it is a good thing to have had an accident—& if one is to have one, one cant well get off easier than we have.

Maria Burnley Bathoe might be viewed as a traveller, since she spent time in India because of her husband's employment. She evidently came from a family accustomed to living abroad, as her letter shows. In 1843, she married Charles Gubbins of the East Bengal Civil Service, the son of General Joseph Gubbins; he and his three brothers all served in India; one of them, Martin Richard Gubbins, published one of the most famous accounts of the Indian rebellion. Charles retired by 1861 and at some time between then and his death at the end of 1867 changed his surname to Bathoe, his mother's maiden name. Possibly this was because of an inheritance from his mother's family; certainly Maria seems to have been left well off. Maria wrote to Darwin on 25 March 1871, inspired by her reading of *Descent of man*.

Descent of Man

Vol 1. Chap 1. Page 21

"But I have never heard of a man who possessed the least power of erecting his ears".

On this point enquiry might be made in the United States—for my mother—Mrs Hume remembers often to have heard from her father Mr Burnley that in his youth he had seen Red Indians who when listening attentively erected the outer ear; curving it up as one sometimes sees a European do by aid of his hand. I personally remember the same having been told me by my grandmother Mrs Burnley: & I think she named the tribe to which they belonged—but it was not retained by my childish memory & my mother (now 84 years of age) can recall nothing beyond the bare fact—

Mr Burnley was a land owner on the James river near Norfolk in Virginia: & had travelled repeatedly towards the Missisippi thro' country then mostly unsettled, which has since become Kentucky & Tennessee—also in the Carolinas— But he quitted America, never to return, on the declaration of independence in 1776—having passed some previous years in arms on the Royalist side: so his observations must date a century back—

Vol 1. Chap 2. Page 46 & 7—

Two or three cases of Reasoning in animals that have fallen under my observation seem to me curious— if not really worth notice the paper can easily be thrown into the fire—

In 1845 I had a pet Antelope at Panneeput (30 miles North of Dehlie) I had brought it up from a few days old & it was quite at liberty

& used to run about the country: but chiefly spent its time in our grounds & always came at sundown to be fed in a half unroofed mud outhouse where it had been kept in infancy & where it was locked up every night to protect it from jackals— I may say locked up with its own consent: for it always walked in voluntarily & never attempted to run out when we closed the door— It would follow me about like a dog & when it attained maturity its affection was more demonstrative than agreeable, as I was often nearly knocked down & once considerably hurt by its bounding upon me— in 1846 we removed to Kurnaul 30 or 40 miles further N.W—& at first the Antelope was kept tethered, then carefully watched till it appeared quite at home in the new home— However one day he disappeared & the next day my husband (who was Magistrate & Collector) received a report from one of his native officers at Paneeput, that my Antelope, well-known by its scarlet leather collar, had returned at sundown to the Paneeput grounds, walked into the outhouse as usual, there been captured & would be sent back to me.— Exactly the same occurred a second time: after which I kept him much longer tethered & under surveillance & really thought he was quite reconciled to the change—but— again he went off, but not to the Paneeput grounds: which he was never known to re-enter, altho' he was several times seen in the vicinity & altho' every effort was made to entice him back again—

Now was not that reasoning?

"I wish to remain near Paneeput—but if I go into those grounds I find that I somehow get removed to Kurnaul—therefore—I will not re-enter those grounds tho' formerly my most favorite haunt"—

In 1850–51 at Meerut 30 miles NE of Dehli, I had a Hog deer, similarly brought up from infancy & similarly running quite loose— Unlike the Antelope however it seldom if ever went out of the grounds, & instead of being shut up at night it lay on its own mat under a portico before which a native sentry paced— It liked to be patted & noticed by us: & it distinctly recognized persons who were much about the house—but its affection was chiefly given to the 4 or 5 men who took turns as sentry (NB Its evening meal was given under this same portico) One day I heard an outcry of "The hunt is coming thro' the grounds" & in alarm for my hog deer ran to the portico— He had been at the x when the hunt came over the wall at & instead of running away as would have seemed natural, he rushed across its course by the dotted line (very narrowly escaping the foremost dogs) to the sentry under the portico— & no farther—but putting his muzzle in the man's hand stood firm while

some of the dogs swerved round towards him— When I knelt down & put my arms round him as a protection & encouragement (while the servants kept off the dogs) he was trembling like an aspen leaf & watching the movements of the dogs in evident terror—but he never attempted to run away from them any farther, nor ever moved his muzzle from the hand of the sentry—

Must it not have been some process of reasoning that taught him to seek the protection of his human friend & comrade? altho' in so doing he actually had to meet the cause of his alarm—

Experience could have no influence: as the only dogs he knew had grown up with him & never molested him: he had never before seen a pack of hounds— Nei[*ther*] are hog deer (in that country at least) ever hunted—as they frequent broken ground covered with brush wood & grasses 6 feet high where riding would be impossible, at any speed—

About this same time & at the same place I had a pet Mongoose: of the small-variety well known (in India) for their susceptibility to domestication: & their habits of extreme personal familiarity— Two were brought to me when 2 or 3 days old: & they were extremely troublesome as long as they required milk-feeding: for they very early possessed little sharp teeth which they employed much more freely than was at all agreeable— After being kept in a box at first, then tethered to a table & afterwards kept loose in an unused room, they were left free to run about the house which they kept quite clear of snakes & other vermin— For some months they never were known to quit the house: but, one never became so tame as the other: that is, it seldom came near us save at feeding time, & tho' it w^d obey a call & allow itself to be handled it did not court such notice— Both markedly preferred my husband & myself to any natives—even the servants who assisted in feeding them— indeed they would peep into his business room & if the native clerks & attendants were there in numbers would immediately retire; altho' they ran about there as freely as in any other room out of office hours— (this may have been smell).

When we were absent for some months in camp they apparently less liked their home—& on our return we found they were a great deal in the garden (where were plenty of their fellow-creatures) only coming into the house for food—& one soon ran wild altogether— The other remained perfectly familiar—would play about us the prettiest tricks, especially a kind of hide & seek, delighting in a chase—or it would run up onto our laps: hide in his loose sleeves, or behind his waistcoat—or run over my shoulder & drop down on my lap rolling on its back & inviting play & tickling like a kitten— It no longer used its teeth so as to give pain, but would seize a finger in its mouth & mumble it by way of caress: & the only mischief it ever did, was to the muslin frills & lace trimmings of my dresses: which it would tear into ribbons with its teeth while holding them down with its paws.

So far I believe it only resembled many others of its species who are often kept as pets especially by young men—

But one of its habits was I believe unique— After it frequented the garden it no longer required regular feeding: sometimes it got nothing from us for days together, still running about the house & playing with us as usual—but when it was hungry it wl^d. come straight up to the sofa where I (being in ill health) was generally lying & stand up on its hind legs to beg like a dog—wh: we certainly never taught it to do— Often I tried the effect of not noticing the entreaty— It would then run up onto me, & in some way or other attract my attention, which being accomplished it would suddenly leap (not run) on to the ground & again stand up to beg— When however I rose from the sofa it did not follow me, but raced away (losing sight of me meanwhile) straight to a side table in the dining room, where under a heavy cover was a dish with shreds of meat always ready— Here I would find it waiting & however slowly I might move, it waited quite patiently so long as I continued moving but if after reaching the table I delayed more than a reasonable time in beginning to feed it, up went the little animal again in the begging posture: & once when I intentionally delayed a long time, it somehow scrambled up the skirt of my dress on to the table & there set itself up under my eyes—

Could it more distinctly say, I am hungry— attend to me—

I never knew it thus beg without being ready to eat whatever I would give—tho' it often refused food that I offered at other times—& altho' its subsequent proceedings showed how well it was aware of the locus of the food, it never went there till after it had secured my presence: which surely exhibits a connexion of ideas not distinguishable from reasoning—

We left the place suddenly from severe illness & had no means of taking our little pet who soon ran wild again I dare say.

In no one of these cases can the principles of heredity, account for any of this mental cultivation— all were the offspring of perfectly wild parents (& grandparents ad infinitum) who had never been domesticated or perhaps seen a human being—so the circumstances in which they were placed were totally novel—

Vol 1. Chap IV— Page 142. note— "Dr Bachner xx has given good cases of the use of the foot as a prehensile organ by man":

Do these include the carpenters & tailors of Upper India who habitually hold one end of a long piece of wood, or a long seam on which they are working between the great toe & its next neighbour— indeed I have seen one of my native maids, pick up a light thing from the ground with her foot instead of stooping for it: but she may have been an exception as I cannot remember to have seen any other person do so— None of these people were ethnologically of low classes—not of any of the aboriginal tribes, nor the black Tamul races—but of the true Aryan race, delicately formed, with black eyes, long straight black hair & skin a light mahogony color: low-cast Hindoos—

Pray do invent some *name* for our remote predecessors— Some sonorous but not difficult word of classical derivation (like Hipparion)

You are writing for the general public, with whom there is a great deal in a Name (begging the poet's pardon) & the unpleasant, as well as awkward phrase "the ape-like progenitors of Man" excites a very unnecessary degree of prejudice against the System.—

As the facts above stated rest on my personal testimony I feel it needful to sign my name, altho' perfectly aware that it possesses no weight social or scientific—

Maria Burnley Bathoe

6 Bryanston Sq— March 25—

Lady Florence Dixie was the daughter of the eighth marquess of Queensbury. Her father died when she was 3 years old, and at the age of 7 she and her siblings were nearly separated from their mother, who had become a Roman Catholic. She was later to campaign against the lack of rights women had to keep their own children in cases of family disputes. She went to a convent school, which she did not much enjoy, and took to writing poetry. She had a passion for travel and big-game hunting, enthusiasms that she shared with her husband, Sir Alexander Dixie. She was the moving force behind an expedition to South America with her husband and two brothers in 1878. Her account of it, *Across Patagonia*, was published in 1880. Partly as a result of her book, she was appointed the *Morning Post's* war correspondent in South Africa. Later in life she became an advocate of complete sexual equality and campaigned against blood sports, which she had become convinced were cruel; she also supported secularism. She corresponded with Darwin in 1880.

> *Glen Stuart,* | *Annan.* | *N. B.*
> "October 29th. Friday.—"

Dear Sir.—

Whilst reading the other day your very interesting account of "A Naturalist's Voyage round the world"—I came across a passage descriptive at Maldonado of the subterranean habits of the tucutuco in which you express the belief that this animal never comes to the surface of the ground.— I am sure it will be interesting to you to know that tho' this *may* be the usual habits of the tucutuco that there are exceptions. In 1879, I spent 6. months on the Pampas and in the Cordillera Mountains of *Southern Patagonia* and during my wanderings over the plains I have had occaision to notice in places tenanted by the tucutuco, as many as five or six of these little animals at a time outside their burrows. This was on moonlight nights, and I cl[d]. not possibly be mistaken as they w[ld]. frequently come within a yard of the spot on which I lying.— On *two* other occaisions I have seen the tucotuco in broad daylight come out of its burrow and shuffle awkwardly along some 20 or 30. yards ere it took refuge in another of the hundreds of holes with which the ground appeared undermined. On one of these occaisions an indian who was sitting near threw an unfinished stone ball of a bolas which he was fashioning at the animal and killed it.— A dog immediately carried the body off so I was unable to examine it and see whether its eyes appeared blind or not.— The other one which I caught could see well enough & when I let it go shuffled quickly away.— I feel sure you will forgive me writing what I have done but I felt that what I personally saw w^{ld} be interesting to prove that on some occaisions the tucutuco does come to the surface of the ground.—

Trusting you will forgive the seeming presumption on my part I beg to remain | very faithfully yours. | Florence Dixie.

Darwin evidently recommended some books to Dixie.

Bosworth Park. | Hinckley. | Leicestershire.
Nov. 4th.

Dear Mr Darwin.—

I must write a line to thank you for your kind letter in reply to mine.— The books you recommend I shall certainly procure & read with interest;— I have myself written a short description of my wanderings in Patagonia which appears this month in print and if you will do me the honour of accepting a copy I shall feel very proud to send you one. The work does not comprise the extent of my whole expedition which on leaving Patagonia I carried on up the Rivers Plate Uruguay & Parana. From Patagonia I brought home some ostriches a gunaco, & from the Rivers Plate, Uruguay, & Parana, a great many animals, comprising some ostriches, a Capybara & a little jaguar. The mother attacked me & followed me up a tree, in self defence I was obliged to shoot her but saved one of the cubs from the gauchos.— Since then he has been my almost constant companion following me ab^{t}. like a dog altho' of an enormous size being now 2. years old. I only yesterday took him to the Zoological Gardens, much to my regret, but he was growing so big that it was not safe keeping him longer at large. I have mentioned this fact to prove how these animals can be tamed by kindness as completely as a dog.—

With many apologies for thus troubling you | I beg to remain | very faithfully y^{rs}. | Florence Dixie

A little later Dixie sent her own book to Darwin.

Lady Howard Dr

GLEN STUART,
ANNAN,
N.B.

" October 29th. Friday:—

Dear Sir.. Whilst read=
=ing the other day your
very interesting account
of "A Naturalist's Voyage
round the world" I
came across a pas=
=sage descriptive at
Maldonado of the sub=
=terranean habits of
the Tucutuco in which

Letter from Lady Florence Dixie. DAR 162: 182.
By permission of the Syndics of Cambridge University Library.

Bosworth Park. | Hinckley. | Leicestershire.
Nov^ber. 29th.

Dear M^r. Darwin.

I have great pleasure in forwarding to you the Acc^t. of my travels in Patagonia and trust the book will meet with your approval.—

I fear you will find it devoid of much interest. While begging you to look on this my first literary production with leniency written as it was during the never ending interuptions of a London Season I hope that the hurried acc^t. of a few of our adventures & occupations in that far off land will be of sufficient interest to carry you thro' its pages.— I venture at the same time to send you a little tragedy [*Abel avenged*] which I wrote some years ago as a child of 14.— It was printed for private circulation at the request & after the death of the late L^d. Lytton but having given some offence I have since suppressd its circulation.

With many apologies for what may appear presumptions on my part allow me to remain | yr^s. very sincerely | Florence Dixie.

Marianne North was an indefatigable traveller and an extraordinary artist. She travelled with her father until his death in 1868, when she was 38; after that, armed with introductions from her father's friends, including Joseph Dalton Hooker, the director of Kew Gardens, she travelled alone, with the purpose of painting tropical vegetation. She travelled to Canada, the US, Jamaica, Brazil, Japan, Borneo, Java, Sri Lanka, and India. In 1880, she presented her collection of paintings to Kew, for display in a gallery designed and paid for by herself. In a wonderfully modern spirit, a refreshment room was part of her plan; Hooker declined this amenity, however, citing the difficulties of keeping the British public in order and providing enough food and drink. While this was being settled, Darwin asked her to visit him at Down. North was an instant acolyte of Darwin's: she wrote in her autobiography, 'I was much flattered at his wishing to see me, and when he said he thought I ought not to attempt any representation of the vegetation of the world until I had seen and painted the Australian, which was so unlike that of any other country, I determined to take it as a royal command and to go at once.' She duly travelled in Australia and New Zealand in 1880 and 1881, and after that in South Africa, the Seychelles, and Chile. Regrettably only one letter survives from Darwin and North's correspondence, which may not have been extensive. It was written to North after a visit to Down on her return from Australia in 1881. North describes the scene in her autobiography: 'He sat on the grass under a shady tree, and talked deliciously on every subject to us all for hours together, or turned over and over again the collection of Australian paintings I had brought down for him to see, showing in a few words how much more he knew about the subjects than any one else, myself included, though I had seen them and he had not' (North 1893, 2: 87, 215).

Down, Beckenham, Kent.
2d August 1881.

My dear Miss North,—

I am much obliged for the "Australian Sheep," [*a shrub, Raoulia eximia*] which is very curious. If I had seen it from a yard's distance lying on a table, I would have wagered that it was a coral of the genus Porites.

I am so glad that I have seen your Australian pictures, and it was extremely kind of you to bring them here. To the present time I am often able to call up with considerable vividness scenes in various countries which I have seen, and it is no small pleasure; but my mind in this respect must be a mere barren waste compared with your mind.—

I remain, dear Miss North, yours, truly obliged, | Charles Darwin.

13 Servants and governesses

Servants were a vital part of middle-class Victorian life. Female servants and governesses were part of the great army of women earning their own living. According to Harriet Martineau, writing in 1864, more than two million English women were self-supporting workers (Martineau 1864, p. 554.) When reading Darwin's remarks about how women could not be men's intellectual equals until they were generally breadwinners, it's useful to remember how many female breadwinners were living in his own household. Little correspondence survives between Darwin and his servants, and most is of a strictly businesslike nature. The Darwins were reportedly kind to their staff, who as a result stayed longer than they might have in other households. Emma and the children maintained relationships with some servants long after they had left the family, and where letters from the servants do survive, they tend to have been sent to Emma and the children. From these and letters to and from Darwin himself that mention servants, it is possible to get a clearer idea of their lives.

When the children were tiny, Emma had nurses to help her, the best loved one being Jessie Brodie. Before joining the Darwins, Brodie had been nurse to the children of William Makepeace and Isabella Thackeray. Brodie left Down after Annie's death in 1851, overcome by grief. Darwin paid her an annuity of £5. She retired to Portsoy, Scotland, near her birthplace. She visited the Thackerays and Down regularly until her death in 1873, and kept in touch by letter. She wrote the following letter to Henrietta after Henrietta's marriage to Richard Buckley Litchfield in 1871.

3 Banermill St [*Aberdeen*]
22 Nov

My ever dear Mrs Lichfield

Think how happy I was when I got your very kind letter with that beautifull adress to Mr Lichfeld   it is just what his apperance is in his Cards, where it is Relley Sumthing to be admired   it Shall be quit a treasur to me to look at & admire it. it was so very kind of you my owen sweet pet to send it   I Cannot Describe how much I estame it. it is more than I Could have Expected of you to have sent it   I haup you will Geat a Comfortable House   I Shall hear from Down   you mensoned Dearest not to think alltho you Did not writ me often in the Winter I Could not expect you Could   you will Have to attend to all your Household Deauties but abouve all your Dear Husband may

194

the Lord Spear you both long to Gather   I not going to writ much more at Present it is very Dark and Dissmall to Day I haup all the Dear Family at Down ar Well I am affraid Dear kind Mr Darwin is not making much Progress in helth now

My Sweet Child I Am just going to finish   What I Can I Say to Dear Mr Leachfeeld   Culd I Send Him my love you know I love you Derest & if I do love you I ought to love Him   I haup you will Both excuse me for my shortcummings—

Good by my Dearest one my the blessing of the Lord Rest uppon you and your Dear Husband

I am my Sweet Child | Your ever affect | Nurs Old Brodie

When you read this burn it

When the children were a little older, Emma employed a series of govern-esses. Governesses were generally middle-class girls, of not much lower status than their employers, but needing to earn a living. They could form close relationships with their employers and be passed around families of friends who could be relied upon to find them jobs. Emma kept up life-long relationships with governesses that she liked, for example the Thorleys and the Ludwigs.

The first governess recalled by Henrietta in her reminiscences was Catherine Thorley, and this account of the Thorleys is from her some-what resentful narrative (DAR 246: 31–5). Catherine's father, a solicitor, died leaving a widow, four daughters, and a son. The family were pro-tégés of the Tollets, who were also friends of the Darwins. Three daugh-ters, Catherine, Lizzie, and Emily, worked as teachers or governesses; the fourth was mad. The best of the three, in Henrietta's view, was Lizzie, but she was taken by the Wedgwoods at Barlaston. The sisters earned enough to send their brother to Oxford, where he wasted his time: he never earned any money. Later, the family came into a small inheritance, and although half was lost in bad investments by the brother, they were able to retire on the rest.

Catherine was governess to the Darwins from 1850 to 1856, and was present when Annie Darwin died at Malvern at the age of 10 in 1851. She stayed in Malvern after Darwin's departure to help Fanny Wedgwood and Jessie Brodie to arrange the burial. (For Catherine's letters to Emma, see chapter 3.) Fanny wrote to Emma and Charles after the funeral.

Friday the 25 April

My dearest Emma & Charles—

On this day I must write to you knowing that our hearts have one common feeling, & that I need not fear to hurt you—in tell-ing you that Hensleigh & I are just returned from that sad & last work of laying your dear child in her earthly resting place— We went at 9 o'clock & I think every thing was rightly arranged as you would have wished— I feared for Miss Thorley & still more for poor

Brodie & she has suffered poor thing most sadly & had to be lifted into the carriage, but since she has been relieved by a long fit of crying & is lying down now— Miss Thorley was more composed than I expected only now & then with bursts of grief—poor thing— There never could have been a child laid in the ground with truer sorrow round her than your sweet & happy Annie— How very kind in you my dear Emma to write me your note—at such a time. I felt it such a comfort on my return to the house to find it, & that you were then as I hoped having Charles with you— I cannot be surprised at yr fears for him it will be hardly possible that he should not be ill— I shall be very anxious to hear how he arrived & I know I shall have a line on Monday from dear Fanny [*Allen, Emma's aunt*] or Elizabeth [*Wedgwood, Emma's sister*] I am sure he could not have gone through being with us this morning—& have never ceased being thankful that he went yesterday—

I have advised Brodie to go straight home tomorrow she longs to be there & when she finds you want her I hope she will be able to put restraint on herself She can be at Down before 9 oclock we have taken Miss Thorley out a drive & she is better— She thanks you with her grateful love for your message & she will write to you fm home— We all leave this sad place together at 9 tomorrow morning— Thank you dearest Emma for telling me I had given you some comfort—& Thank God that you are well—

Y^r most affectionate | F E W—

Darwin wrote to Catherine Thorley's mother, Elizabeth, to thank her for Catherine's help.

Down | Farnboro | Kent
Ap 26.

Dear Mrs Thorley

I must beg permission to express to you our deep obligation to your daughter & our most earnest hope that her health may not be injured by her exertions I hope it will not appear presumptuous in me to say that her conduct struck me as throughout quite admirable. I never saw her once [*yield*] to her feelings as long as self restraint & exertion were of any use— her judgment & good sense never failed: her kindness, her devotion to our poor child could hardly have been excelled by that of a mother

Such conduct will, I trust hereafter be in some degree rewarded by the satisfaction your daughter must ever feel when she looks back at her exertions to save & comfort our poor dear dying child. I earnestly hope that her health will be pretty soon established. My wife joins in kindest remembrance to yourself

Pray believe me &c

Mary Ann Pugh was governess from 1857 to early 1859; she seems to have suffered from severe depression. According to Henrietta's reminiscences, she used to sit at meals with tears pouring down her cheeks (DAR 246: 36). She later lived at the Priory, Roehampton, then a mental hospital, and at Kew, where the Darwins' friend Joseph Dalton Hooker was asked to keep an eye on her. Emma apparently paid for her to have a regular holiday (*Emma Darwin* (1904) 2: 261). Towards the end of her employment, Emma wrote to William, 'I certainly feel it a great relief Miss P. going & so does she I think, as she seemed in better spirits. We are not going to have Miss Thorley now tho' I think we may in time, but I am looking out for a Swiss for 6 months or so to set you all talking French & German' (DAR 210.6: 34).

According to Emma's diary (DAR 242), a Mrs Grut was engaged on 24 January 1859, although it's not clear that she was the promised Swiss: Henrietta described her as a 'quite mad Dane' (DAR 246: 38). Mrs Grut proved to be something of a terror. This letter is from Emma to William, written on 23 February 1859:

> I thought at Hartfield that I sh^d have come to a regular blow up with the G. but I wrote her a note "stern but just" (as she talks so much she never listens to a word I say) & that brought her to reason & we have gone on quite smooth since   how long it will last I know not. She used to keep poor Skimp [*Horace*] stupifying over a lesson for 2 hours & then told him he sh^d have no breakfast till he said it   Fancy the consternation of Frank & Lizzy   they came solemnly to tell me, & then she scolded them for telling tales, & then I scolded her & said they always were to tell every thing & then she scolded me & said they were such naughty children behind my back I c^d not judge any thing about them. But we are the best of friends now, but I feel there is a volcano beneath the surface. She gives a German lesson to Etty & I try to listen & make out what she says in German.

Emma wrote to William again on 12 March 1859: 'Mrs G. is on her good behaviour but talks too much when she comes down stairs so that she is tiresome. I expect we shall rub on pretty well now. She laughs so much that I feel my face getting quite rigidly grave' (DAR 210.6: 39).

Finally, in March 1859, the inevitable breach occurred: this time Henrietta filled William in:

> My dear William. Solemn events have happened. Mrs Grut is gone for ever. this is how it came about. On Monday at breakfast Mama said very civilly that she wanted some alteratiration in Horace's lessons. Mrs Grut was evidently huffed at that, & then I said that I thought s'eloigner wasn't to ramble very mildly & that huffed her again & she made me some rude speech or other "Oh very well if I knew better than the dictionary," I quite lost that speech then, Ma repeated it afterwards.

Nothing more came of it then, & all went smooth till I went up to my German lesson in the evening. When I came in I saw there was the devil in her face, well she scolded the children a bit & then sat down by me, when I showed her my lesson (a bit of very bad french) she said, if I knew better than she did it was no use her teaching me & so & so on, till it came to a crisis, & she worked into a regular rage. Oh *you've* no feelings, but I have, I feel these things & so on.

This all referred to my saying I thought s'eloigner wasn't to ramble, she had been brooding over it all day & then when she had got me all to herself it burst forth, I left the room then, & went down stairs to tell my injuries.

When Papa & Mama heard all about it they settled she shd go at once, so Papa wrote a letter telling her she shd have her 33 £ & nothing more, Mama went upstairs to watch till the children had done, they all came out crying. After she cdn't scold me she fell upon them I suppose, then Papa was to go upstairs & deliver the letter. The very first thing she said was "*I* don't care for your legal notice to quit" & tossed the letter on one side, "Mrs Darwin engaged me for six months & you'll have to keep me. The line of conduct they had settled upon was not tell her why she was sent away so that if there was a lawsuit it might only rest upon whether we had the right to dismiss her without paying her board & lodging. Papa got *such* a torrent, telling him he was no gentleman, & white with passion all the time. Wanting to know what she had done, what he had to accuse her of— telling him he was in a passion—she would give him time to think.

Some time ago Papa wrote to his solicitor, to ask him what he was to to do about salary & his advice was pay the salary but not the board & lodging. We had a very flustered tea, & all evening we sat preparing for the worst, what we shd do if she refused to go out of the house etc. However she did turn out much milder & sent us a letter to say she wd go on Wednesday.

According to Emma's diary, Mrs Grut went on 16 March; the row had taken place on 14 March. Emma may have recalled Miss Pugh in the emergency, as she wrote to William on 24 March 1859.

Down Thursday

My dear William

It is a long time since I have written to you but I have been so busy teaching since the departure of the G. Etty undertook one child which helped me much.

Miss Pugh came on Tuesday for a short visit before settling with her pupils. The children are very glad to have her again & if she was always as cheerful & pleasant as she is now, we could not desire any thing better. She has the children to lessons & teaches them dancing. She saw Mr Hawkshaw in London so that he means to remain friendly with her but I don't suppose it will ever be made up with M^{rs} H.

…

Papa has been expecting a lawyers letter from the Grut but I think we shall escape now.

The problem with Mrs Hawkshaw may have been the one suggested by Henrietta: Miss Pugh thought Mr Hawkshaw was in love with her; or possibly the other way round (DAR 246: 36–7).

Miss Pugh visited again in June 1861; Emma wrote to William on 3 June, 'We expect Miss Pugh on Thursday for 2 nights. She has left the American family who treated her *dreadfully* by her accounts but she gives no particulars' (DAR 219.1: 46).

Catherine Thorley's sister Emily took over as governess while the Darwin children were at Ilkley in Yorkshire, where Darwin was receiving the water treatment, in October 1859. Emma described the situation to William.

We have got Parslow & two maids & a comfortable house & Emily Thorley. I am quite awfully well & strong & eat immensely as we all do, & the mutton is something peculiar. We have had bitter cold but I am glad to see the rain tonight, tho' we enjoyed a long walk up the hill in the frost today & along a rushing brook with little waterfalls. Frank found a dead grouse to his great delight. It seemed quite fat & fresh, so no doubt it was shot & we mean to cook it   We almost killed poor little Emily with the walk.

We have a small supply of lessons which saves much ennui. I wish we had Miss Thorley for the walking sake & Lizzy, but Emily is the pleasantest of the two. We shall certainly stay a month longer & your father says he will stay till Xmas if it keeps on doing him good. His stomach is quite wonderful certainly. …

Etty is well & I have some hopes that it may give her some vigour but there is no difference in her walking as yet. Goodbye my dear old man. I am afraid John Thorley is thoroughly idle. His sisters seem very fearful of his not passing. It is too bad of him when such an effort was made by his mother to send him to college. | Yours E.D.

Emily kept up a correspondence with Henrietta and Elizabeth; this letter is from 1864. Her tone is indistinguishable from that of a sister or cousin, and her social activities are very much the same as those of her former pupils.

Wednesday

My dear Henrietta,

It seems a long time since I wrote to you, but I can't remember how long. Please do say in yr next how you mama is; in yr last to me she was so far from well. I am sorry to hear Horace does not get on very fast poor little man.

What a gap Lenny must have left in the house. Thank Lizzy for her nice letter, & say I will write to her as soon as ever I have got any thing to say.

We are going (all three of us) to yr Aunt Fanny's on Saturday for a singing practice. We shall set to work at 8 o'clock. How I wish the Fates wd decree that you should be in town that day!

Have you seen Holman Hunt's "Light of the World"? In some respects I like it much better than Christ in the Temple, which I remember you have seen. I think I have never seen such eyes on canvas (in the former I mean). I declare you cd almost imagine there must be a soul to give them such expression & life. The whole picture is such a beautiful allegory—not the smallest object in the work but has a symbolic meaning. It is put by now, but will appear again at the Grt Exhibition.

I should think your cousin Geoffrey had made an excellent choice of a wife. I feel to know her pretty well, for we had a good talk together at Barlaston. By the way I met Holman Hunt the other night, & am sorry to tell you that he *talks* vulgarly. His countenance has a very *good* expression, but it did not strike me that there was that intellect & *refinedness* in it that I looked for.

You asked me how I liked that Sonata of Beethoven "Les adieux, l'absence & le retour". To say that I liked it is much weaker language than I mean to use— I declare some parts of it transported me to sixth if not the seventh heaven—some glorious bits seemed actually to *force* one to smile. I shd. like you to get well acquainted with that passage, I think it is l'absence leading to le retour, & on through that movement— if you don't like it I believe my name is not Emily. But I cannot write you half what I think of it. What a little darling Quiz must be— how you will all love him. I am *trying* to learn "Adelaide"— with what success time will tell. Isn't this biting weather.

Your's ever affectly | Emily Thorley

I hope you have not wanted yr Molière— I am afraid I'm keeping it too long. The very week before we were to have acted "Le Bourgeois Gentilhomme" two of our best actors lost near relations, so it is put off till the Spring. We had good fun at the rehearsal.

From 1860 to 1863, Camilla Ludwig of Hamburg was governess; Emma kept in touch with her long afterwards. She helped Darwin with German translations. The year that this letter was written is unknown, but it must have been while Ludwig was away from Down or after her employment there; the German translation in this case was evidently rather mundane.

Down
Feb. 21

My dear Camilla

Will you be so very kind as to look at the enclosed letter sent me by some old & goodhearted man.— No one here, visitors & all, can make it out.— *I do not want Translation of whole*, but only the part in which he recommends some treatment. I doubt, however, whether I shall try it.— Does the writer seem to know what is the matter with me?—

Yours affect | C. Darwin

On her return to Down in November 1862 after visiting her family in Hamburg, Ludwig reported that her family was in great poverty and that she had advanced her whole salary to them (DAR 219.1: 64). Darwin's Account book–banking account (Down House MS) records a payment to Miss Ludwig of £200 on 21 December 1862. Next to the entry Darwin wrote '(Class as Invested)'. On 22 December, he wrote to Camilla.

Down

In all money affairs it is better to have a memorandum in writing. I thought it so hard a case that your Father should be told that the £200 would be given him & then the offer partially retracted, that I, under these circumstances alone, offered to give the £200, which I have now done: but I understand that if Miss Davies kept to her original intention, then, as I know perfectly well, you would repay me the £200; under any other circumstances, I must freely & cordially give you the £200. This note is ill expressed but is written in a hurry.

My dear Miss Ludwig | yours very sincerely | Ch. Darwin

The children had different opinions of Camilla. Henrietta described her as 'a very incompetent German governess, who taught us no German & not much of anything else' (DAR 246: 38). Horace became devoted to Camilla; Emma wrote to William on 2 March 1862: 'Horace's devotion to Miss L. is got to such a pitch that I don't know what he will come to. He can't bear to sit on diff' side of the table at meals so that he often gives up the fire side for the sake of sitting by her' (DAR 219.1: 49). On the advice of the village doctor, Stephen Engleheart, the decision was made to try to separate them; Emma wrote to William on 27 May about a way she had thought of to do it that would minimise Horace's distress. Louisa Ludwig was Camilla's sister, who took over at Down while Camilla had an extended holiday in Hamburg. 'Bad' means ill rather than wicked; Horace had been in poor health.

Down
Tuesday

My dear William

A sudden flash has come into my mind that I will bring Horace & stay a week with you. Miss Ludwig is going to Hamburg & we think it will break his heart much less to leave her here & come to you which he w^d enjoy so much than for her to leave him quiet at home. I must bring Louisa L. I am sorry to say as she is much more used to him when he is bad than a maid w^d be. I think you have two spare rooms & I wonder whether you c^d hire a small bed or big crib to put into one of them. We shall not bring a maid. We should like to come this day week if that will do for you & Miss Pugh will stay at home with Lizzy. Mr Engleheart is very anxious to get him away from Miss L. & he thinks a change or excitement would do him good. I doubt whether it will make much difference about Miss L. but I think some times his great fondness for her agitates him & makes him worse. I think she is very judicious & quiet with him.

Unfortunately, in June, Leonard came down with scarlet fever, and while the Darwins were taking him to Bournemouth to convalesce, Emma too became ill, and they all had to stay in Southampton with William. Darwin replied to a letter from Camilla, still in Hamburg.

1. Carlton Terrace | Southampton
Aug. 26th.

My dear Miss Ludwig

We thank you sincerely for your heart-felt sympathy. We have had a bad time of it; but Emma is nicely recovering, & came down to the drawing room for a little bit to day. She had considerable fever. Leonard travelled too soon, & was injured by the journey. Poor little fellow, he has been so patient during all his terrible illness; he will be months before he will be strong. We hope to move to Bournemouth in about a week, if both patients go on well. Elizabeth Wedgwood is here, & has been, as usual, the most unselfish & devoted of nurses.—

Horace is going on well & only occasionally has a baddish day. Etty is accustoming him to have no one to sit with him at night; & she has so much judgment & kindness, that she will do it well. It is a horrid bore, but we have been forced to engage a second house at Bourne-mouth, & so shall not be all together.— Emma sends her love & will write when she is strong: she hopes that Horace will soon be able to do some lessons, & then it will be capital for him to have you to return. But I hope we shall then have got him out of his invalid habits. Poor little man he has often cried, when he has tried & failed to write to you. And no wonder for nothing could have possibly exceeded your kindness to him.— What a wretched summer & spring we have had! We had a very nice letter lately from Louisa, written with so much feeling. I fear some things are very uncomfortable at the school; but she seems determined to bear them with excellent spirit. I sincerely hope that you have been happy & enjoyed yourself at home with as few drawbacks as this weary world permits.

My dear Miss Ludwig—
Yours very sincerely | C. Darwin

Camilla returned later in the year and Emma wrote to William on 6 November.

> We were expect Camilla all Tuesday & at 12 o'clock at night Hen. who was awake heard a fly drive to the door & then to her surprize nothing further took place & the fly drove off so she came & waked us up & then Anne [*a maid*] tried in vain to wake James [*a footman*] & send him down at last poor Miss L. & her boxes were found in the yard as she did not like rousing the house with the door bell. Her plan of not disturbing the house did not answer very well. The next day I had to tell her that Lizzy & Horace were going to school & it was a terrible blow to her poor thing. She has had a sad visit at home as I am afraid the family are in great poverty & she has advanced her whole salary for them. Horace sticks to her very much but not in that fidgetty way. He does his lessons very well.

On 19 November, she wrote again.

> It is most extraordinary but Horace has never been so well or cheerful since Miss L. returned. We are going to try her going on a visit to the Malthuses & if he is better & more stout hearted in her absence she must go poor soul as it will be so very bad for his going to school if he is not at his best. Lenny is not very well either. It is mesmerism about Miss L. for it is nothing in her manner or ways. It was very painful telling her about it & that she w^d probably have to go sooner but she took it very nicely & quietly.

In 1874, Camilla married Reginald Saint Pattrick, the vicar of Sellinge, Kent. Her sister Louisa came to live with them. She and Louisa continued to visit the Darwins, in due course with Camilla's own children. Camilla seems still to have been in straitened circumstances, though, as she had only one maidservant. Emma wrote to William on 14 September 1883:

> Yesterday I had Louisa Ludwig & the 2 little Pattricks. They are most lively jolly children, & the chatter & noise are ceaseless— I am glad to see to they eat a reasonable quantity now— Louisa's soul is entirely concentrated in them & she can speak of nothing else, which I don't at all dislike— They only keep one maid now, Camilla does most of the cooking & Louisa some house work— The tithes were only £180 last year; but I dare say Mr Pattrick does not get his dues, as he can't bear struggling for them.

A later letter from Emma to George, on 6 September 1893, reveals that Emma passed on her dresses to Louisa: 'Louisa Ludwig spent a day here in her way to Hamburg— I was quite dazzled by her appearance in a well made grey poplin & I learnt that I had given it to her in 1882— I thought I remembered all my best gowns, but I had forgotten this' (DAR 210.3: 200–1).

By and large the servants in the house were managed by Emma with the help of Joseph Parslow, the butler, which explains why Darwin writes relatively little about them: he might be called on in emergencies, but they didn't take up much of his time. For a flavour of how time-consuming the servant problem could be for a husband if his wife wasn't well enough to do the job, it's interesting to look at this series of letters to Darwin from his friend, Joseph Dalton Hooker. The first was written on 5 May 1862.

> Kew
> Monday
>
> Dear Darwin
> …
>
> We have had a regular kick up, & been in a troublous state for some time. We have had **all** our spoons forks &c to the tune of 80 pieces of silver walked off with; by a nice young man who introduced himself to our maids, & was made so much of that he could not make enough of us without & I have had tears, groans, hysterics, Police inspectors & all the other evidences of civilization in the house. It is all our own faults, wholly & entirely, for not looking better after our servts, doors & establishment. I don't care a brass farthing for the silver, which I have replaced already (with electro-plate!) but I must confess to a feeling of shame at finding out that my establishment has for some months had the reputation of being not a little **disreputable**—

> *Royal Gardens Kew* | Kew
> June 9/62
>
> My dear Darwin
> …
>
> We are still in domestic perplexity— My wife is very thin & watery, lacks energy, blood & muscle. & though she does her best honestly & heartily with the children, she lacks energy & method & does not get on.— We have agreed to a plan of housekeeping which will I hope answer better—to get a middle aged cook who will be sort of housekeeper in as far as keeping an absolute control over who comes in & out of the kitchen— Mrs Darwin is not likely to know of such a person, but perhaps you will kindly mention it to her—quite a plain cook is all we want, who can roast, boil & bake, but she must be beyond the age of flirtation— I can promise her a quiet place, a most indulgent mistress & good wages.— Then we shall return to the Nursery Governess plan—but endeavour to get an older & more practised one than we had before—

I am getting over my dispiriting feelings of annoyance & anxiety. I wish I could add that my wife was better— she complains of palpitation of heart & shortness of breath & she has hardly a perceptible pulse—that she looks very ill every one says. I want above all to take her away, but neither of us can leave home till our household is arranged.

Ever yrs affec | J D Hooker

The story continued on 19 June:

> *Royal Gardens Kew* | Kew
> 19th.

D^r D^{rn}.
…

We are in statu quo—no Cook yet, though we hope we have heard of one who will suit— We have also I hope found the right sort of person for Nursery Governess— Meanwhile we are carrying on the war as best we can & have a party of 12 persons chiefly Italian Botanists coming to a Tea dinner tomorrow.

> *Royal Gardens Kew*
> June 28/62

My dear Darwin
…

I am still in perplexity— We have found no cook yet at all to suit us & I have determined to send the children to Worthing with the Governess, who seems a capital person, & if possible take my wife to Switzerland for a fortnight— She seems to go down hill steadily, & complains of shortness of breathing & palpitation of heart— Whether it is all weakness (as the Doctors say) or symptoms of the affliction her father died of, God only knows, & I dare not ask myself— I *must* get her away but my Father is again laid up with Eczema on the legs, & I do not see how I am to go.

One begins to wonder whether the Victorian habit of sending middle-class invalids to travel abroad was as much to do with getting them away from responsibility for their own servants as with change of scene.

> Kew
> July 2/62

My dear Darwin
…

We have got a Cook, who I hope will suit, she was Sir F. Palgrave's, during all his widowerhood & is a most trustworthy person, too old, but that is a fault on the right side in our case: & at any rate she will enable me to get things belowstairs put on a proper footing— you cannot conceive the relief it is to us to have found a suitable person! Children & Governess go to Worthing tomorrow—we on Friday to Dover, we shall travel slowly. I also fear length &c of journey to Switzerland, but her heart is there, & just see what weather we have here! I saw Dr Walshe yesterday who examined her & assures me she has no organic complaint & that it is all functional

The respite was brief: by November, Hooker's correspondence was getting in a muddle.

Nov 20th | I send letter written 4 days ago

…

Our *admirable* (perfect) governess is so ill with Lungs, that we must part with her, & she has no home poor thing— My wife is in pack of troubles, I take her today to get tooth pulled, a worse affair for me than for her! for she behaves *shockingly* ill; & I am afraid to have her chloroformed   indeed I doubt if dentist will do it—as I must tell him her heart action is not what it should be.

Yours perplexed | J D Hooker

Two years later, things were no better: Hooker wrote on 15 June 1864, 'We got a new governess & packed her off in 2 days.— my wife is most unfortunate, but I do not think she has any idea of the trouble it takes to get a really good person. I want extremely to take her to the country, but do not see my way to it' (DAR 101: 227–8).

Three years later it was the same story.

Kew
Oct. 18/67.

Dear old Darwin

I find myself obliged to put off half my visit to you. The Governess has gone away for her holiday, & my wife finds the nurse not to her trust, & what with this & the boys home for the holidays, she wants more help in the house, so I have to go to Bath on Saturday to bring up a friend of her's, who will help her. Moreover she does not like the nurse, with reason I think, & wants to get rid of her, but has not grounds enough for a dismissal; & as I think it not safe to let the nurse have the child after she has warning, I have advised her to take the child to Hastings for change of air, leaving nurse here, & writing up to dismiss her from thence— This is an ugly ruse to play, but it must be done & a prudent friend in the house will be a great help to her. The whole thing is a horrid bore ...

Ever aff | J D Hooker

Others among the Darwins' friends had their problems. On 22 March 1867, Henrietta Huxley wrote to Emma, 'I have just got a new nurse—& today she has refused to wheel the perambulator— Is it not provoking. I have not yet got an under nurse' (DAR 166: 287). Evidently wheeling the perambulator was the under-nurse's job, if you had one, or were expected to have one. It seems that Emma, in rural Kent, did not have that problem, at least.

The sexual morality of servants was another bother. One of the rare occasions when Darwin wrote about this was in connection with a badly behaved curate, John Warburton Robinson. Darwin wrote to the absentee vicar, John Brodie Innes, in 1868.

> Down. | Bromley. | Kent. S.E.
> Dec. 10
>
> Dear Innes,
>
> ...
>
> As I am writing I had perhaps add that rumours are very common in our village about Mr Robinson walking with girls at night.— I did not mention them before, because I had not even moderately good authority; but my wife found Mrs Allen very indignant about Mr R.^{s} conduct with one of her maids.
>
> I do not believe that there is any evidence of actual immorality. As I repeat only second hand my name must not be mentioned.— Our maids tell my wife that they do not believe that hardly anyone will go to Church now that Mr R. has returned.
>
> What a plague this Parish does give you.—
>
> Dear Innes | Very sincerely yours | C. Darwin

In the 1870s, as Darwin's correspondence became unmanageable, the Darwins began to think of employing a secretary. In the event, Francis Darwin gave up his medical career and took up the post, but before that they tried out Virginia Isitt, who had been governess to the children of Emily Jesse, Alfred Tennyson's sister. There is a draft of Emma's letter to her, written before 17 September 1871.

> My dear Miss I.
>
> I have received from my niece [*possibly Snow Wedgwood*] your letter to her (in which you say you w^d like to undertake the situation of sec. to Mr D. your testimonials are most satisfactory & I will carefully return them to you—
>
> It is so very doubtful Mr Darwin being able to dictate to his own satisfaction or employ a sec. that I think the best plan would be if you wd come & spend a few days with us that you may judge you may see whether the whole thing & accommodations in the village w^d suit you; & if so Mr D. w^d be glad to make the experiment for about a month.

If you will appoint the 1ˢᵗ day next week that will suit you we will meet you at W. Croydon Station at any hour of the afternoon that you will fix— Mʳ Darwin is so m. of an invalid that I will not ask you to spend the evening w. us but I hope you will

Emma wrote to Henrietta on 17 September:

> Mʳ Powell [*the vicar*] v. goodnaturedly hunted all the village for lodgings for Miss Isitt, & nobody will have her but Mʳˢ Martin, they all say that they do not like to undertake a young lady they require so m. attendence. But that is an affair of money so I don't despair of their changing their minds F is going to prepare some straight-forward dictation for her so that he may not be too m. frightened. We send for her to Croydon Ó [*tomorrow*]. Her last Mʳˢ is a Mʳˢ Jesse (late Miss Tennyson who was to have married Arthur Hallam) I dare say R. [*Richard Buckley Litchfield, Henrietta's husband*] knows her.
> …
> We put Miss Isitt into No 4 for the present. After a time I shall turn her out and shall so enjoy making it nice & comf. for you.
> We are greatly puzzled where to establish her, but I think which-ever bedroom we give her as sitting room we must put up a bell from the study. We have many schemes, one is taking ½ of the tool room.
> It is milking the pencil makes this look so ugly. I am going down this mg to see if I can soften any ones heart to take in Miss I.
> yours my dearest | E.D.

Isitt stayed at Down from 18 to 20 September 1871, but evidently the exper-iment was not a success. She already had a French State Certificate in teaching, and became the first headmistress of the Collegiate School for Girls in South Africa in 1874.

Sacking female servants was usually the wife's job, as Hooker's letter to Darwin about his wife's manoeuvres with an unwanted nurse makes clear. Henrietta, married and living in London, without her mother's access to well-known village girls who could be kept under surveillance by a host of neighbours, no doubt had more trouble than her mother did finding and keeping servants. After sacking a cook, she had to go and recover at Down. This is an extract from a letter to Leonard Darwin of 11 August 1874.

> I felt a great wish for rest here for we'd had such a horrid week before. I found out my cook had gone utterly to the bad & had to be turned out at a moments notice— She went off pretty quietly—but I did feel so thankful to have had a man in the house. In fact when I'm a widow & have to turn a great strong woman out, I'll always have a policeman ready. I'd sooner turn a man away—for you feel as if a woman was so much less civilized & might take to scratching

you or breaking your china— It was a horrid thing to happen & has made me feel that there wd be a good deal to be said for Hotel life—

14 Ascent of woman

Only a few women wrote to Darwin about overtly feminist topics, but many of his female correspondents were involved with the suffrage movement and the promotion of women's education. Often they were also involved in the campaign against vivisection; for some, as for Frances Power Cobbe, female emancipation and protection of animals went hand in hand, both women and animals suffering under a malign social order. Such campaigners often felt that Darwinian theory was on their side. Darwin's books had stressed the continuity of humans with the rest of the animal kingdom, so that it was not feasible for Darwinians to see animals as soulless machines that only appeared to feel pain, as Cartesian philosophy suggested. Also, whatever his own political and personal preferences, his account of female subordination made it seem contingent upon historical circumstances: it could, in theory, be changed.

It's also clear from family letters that women in Darwin's family, and their friends, were very aware of contemporary debates about feminism. Some of them were involved in the setting up of Newnham and Girton Colleges at Cambridge. Henrietta found herself smoking cigarettes with the Stansfelds, well-known feminist campaigners, in France. She had discussions with her friends about what they really wanted from education: was what men had really the ideal? Women in their circle, even without raising any particular banner, were extraordinarily active: they learnt mathematics and physics; they hired tutors; they took examinations; they watched debates in the House of Commons from the ladies' gallery; they attended university lectures if they were open to women.

Darwin himself was reticent on the subject. However, the surprisingly effective combination of women and opposition to vivisection spurred him to become a pragmatic supporter of scientific education for women. Women, with the moral authority resulting from their subordination and their motherly role, were effective advocates for an anti-vivisection law, even without the vote. Darwin lamented that if only they understood the medical benefits of physiological research, they would take a more moderate position. When the tricky subject arose of whether girls should learn physiology, he said they certainly should, if they wanted to. Possibly the drip-feed of barely voiced feminism that he had been receiving from his correspondence also had an effect.

Lydia Becker corresponded with Darwin chiefly about botany (see chapter 5); her interest in women's suffrage was aroused by John Stuart

Mill's petition for female suffrage to the House of Commons in 1866. For a time, her interest in science and her interest in women's position in society marched hand in hand; later, the suffrage movement absorbed all her energies. In 1867, she founded the Manchester Ladies' Literary Society. The two other Manchester scientific societies, the Manchester Literary and Philosophical Society, and the Manchester Scientific Students' Society, did not admit women. As she said in her inaugural address (Blackburn 1902, p. 33):

> They do not throw open such opportunities as they afford for acquiring knowledge, freely to all who desire it; they draw an arbitrary line among scientific students, and say to our half of the human race—you shall not enter into the advantages we have to offer—you shall not enjoy the facilities we possess of cultivating the faculties and tastes with which you may be endowed; and should any of you, in spite of this discouragement, reach such a measure of attainments as would entitle one of us to look for the honour of membership or fellowship in any learned body, we will not, by conferring such distinctions upon any of you, recognize your right to occupy your minds with such matters at all.

But it would not be entirely correct to suggest that for Becker, to study botany was an act of revolution. In her view, science provided a model of rational and calm investigation and sharing of views, and a refuge for oppressed minds. She used Darwin as her example: the botanical paper he had sent her to use for the first meeting of her society had been worked up when he was confined to a sickroom. She wrote to Darwin to ask him for his help two years after sending him her *Botany for novices.*

10 Grove st | Ardwick. Manchester
Dec. 22. 1866.

My dear Mʳ Darwin

Before proceeding to the object of my letter I must try to recall my name to your recollection. I scarcely dare flatter myself, that I can do this successfully, though the remembrance of your great kindness and courtesy to me will never fade from my mind and is a constant source of pride and pleasure.

In the summer of 1863 I ventured to send you some flowers of *Lychnis diurna* which seemed to present some curious characteristics, and though they proved on examination not to possess the interest you at first thought they might have with respect to your own investigations you were good enough to write me several notes about them. You also did me the honour to send me a copy of a paper you had read to the Linnæan Society on two forms in the genus *Linum* and I had the greatest pleasure in immediately procuring a pot of seedlings of crimson flax—and watching the appearances you had recorded

...

I have not been able to pursue my study of the Lychis flowers nor my endeavours to penetrate the mystery of their alteration in form, for since then we have ceased to reside in the country and now, surrounded by acres of bricks and mortar—and an atmosphere laden with coal smoke, I have no opportunity of watching living plants.

But living in a town has its advantages, among others it makes possible such societies as that indicated in the circular I have taken the liberty to enclose. A few ladies have joined together hoping for much pleasure and instruction from their little society, which is quite in its infancy and needs a helping hand. Am I altogether too presumptuous in seeking this help from you? Our petition is—would you be so very good as to send us a paper to be read at our first meeting Of course we are not so unreasonable as to desire that you should write anything specially for us, but I think it possible you may have by you a copy of some paper such as that on the *Linum* which you have communicated to the learned societies but which is unknown and inaccessible to us unless through your kindness. In your paper on the *Linum* you mention your experiments on *Primula* which greatly excite my interest and curiosity, for last spring as I was gathering primroses I was forcibly struck with the difference between the "pin eyed" flowers, and those in which the stigma was concealed beneath the anthers. I have known of this difference in the Polyanthus from childhood, but not until I read your paper was I aware of its interest or importance and now I have just enough information to excite and tantalise, but not to satisfy, a strong desire for more. If you will pardon the presumption of the request, I would beg that your goodness might prompt you to send something you may have on hand in the form of pamphet or paper which would help us to learn the meaning of these curious differences in the flowers, and as we may all hope during the coming spring for the pleasure of luxuriating on a primrose bank we should indeed be grateful for the kindness that had guided us to look more closely into the beautiful things we were enjoying.

I send this with much misgiving lest you may be displeased at the liberty I have taken   if I have a hope of pardon it rests entirely on your goodness.

Believe me always | yours much obliged | Lydia E. Becker.

After the meeting Becker returned the papers Darwin had sent, 'Climbing plants' and 'Three forms of *Lythrum salicaria*'.

Manchester Ladies' Literary Society. | 10 Grove s^t | Ardwick
Feb. 6. 1867.

My dear Sir

I return you—with more thanks than I know how to express, the two papers which you were so good as to entrust to my care. Will

Lydia Becker. From the *Graphic*, 10 January 1874, p. 44.
By permission of the Syndics of Cambridge University Library.

you have the kindness to cause me to be informed of their arrival—
having once lost a book-post packet I shall feel a little anxious till I
hear they are again in your hands—and this induces me to give you
the little extra trouble involved in registering the packet—for which
I must apologise.

I have transcribed portions of them, and made large copies of the
diagrams— I hope this was not wrong—without your permission,
but I thought, as they were printed—I might do so without impro-
priety.

The arrangements in *Lythrum* are indeed most marvellous. It sets
one wondering whether different sized stamens in the same flower
can ever be quite without meaning, and if there is any difference in
the action of the pollen of the long and short stamens in didynamous
and tetradynamous flowers. In the N. O. *Geraniaceae* it seems as if
there might be some transition going on—for in *Geranium* each alternate stamen is smaller, and in the allied genus *Erodium* the alternate
stamens have become sterile. Can it be possible that this genus was
once dimorphic, and one of the female forms having by any means
become exterminated, the corresponding set of stamens have shed
away? If one of the forms of Lythrum were to disappear—two sets
of stamens would be made useless to the species, and it is conceivable
that they might then gradually become abortive.

I obeyed your directions about the paper on Climbing Plants and
the insight into their extraordinary and regular movements was a
new revelation to all of us. I made large copies of the diagrams and

dived into my herbarium for specimens of each class of climbers, bringing up enough to make a goodly show. Luckily a collection of ferns from the islands of the South Pacific recently presented to me contained a specimen of one named in your paper *Lygodium scandens*. Till I read it I had never dreamed of twiners in this class, as none of our British ferns have the habit, but as the "march of intellect" seems to be the order of the day, even in the vegetable world, there is no telling what they may accomplish in time!

Our society appears likely to prosper beyond my expectations   the countenance you have afforded has been of wonderful service, and I do hope that by becoming useful to its members it may prove in some degree worthy of the generous encouragement you have given us.

The ladies who had the privilege of listening to the paper desire to express their thanks to you for it, which I hope you will be pleased to accept.

Believe me to be | yours gratefully | Lydia E. Becker.

Becker reused some of her opening address to the society in her 1869 article in the *Contemporary Review*, 'On the study of science by women'. As part of her research, she wrote to the secretaries of scientific societies in London. She confirmed that women were not admitted as members of the Royal Society, the Royal Geographical Society, or the Linnean Society. Women could attend meetings of the Geographical Society as guests of fellows, or, if they were teachers of geography, by a special card of admission. Although women had distinguished themselves as explorers, a women had never been offered a medal or reward by the Geographical Society. The Ethnological Society (on whose behalf Ellen Lubbock had campaigned) was an exception in admitting women as visitors. Mary Somerville was an honorary member of the Royal Astronomical Society. The Royal Horticultural Society admitted women as fellows, but only so that they could visit the gardens. The Geographical Society, but no other major society, allowed a small number of women to come to their soirées as friends of the president; microscopical clubs and the Society of Arts invited women freely to soirées. Becker's apparently not unsympathetic informant added that there seemed to be no bar on women's publishing papers in the transactions of the societies, if they were really good; but as Becker pointed out, women were as a rule not likely to write really good papers if they were debarred from the regular discussions of other enthusiasts. In the next part of her paper, she discussed the Dublin College of Science, which admitted both men and women. Although women seem to have been slightly in the minority in the lectures, they regularly came first or second in the examinations. Many of Darwin's colleagues published in the *Contemporary*, so it's quite possible that he, or Emma, who kept an eye on his press, read this article.

Although not at the forefront of the feminist movement, Emma, Henrietta, and Elizabeth were aware of and intrigued by what was going on.

Many of their friends and acquaintances became involved with women's education. Henry Fawcett was professor of political economy at Cambridge University, and a member of parliament; he shared John Stuart Mill's enthusiasm for feminism. He married Millicent Garrett, sister of Elizabeth Garrett Anderson, the first female physician to train and practise in Britain. Their daughter, Philippa Fawcett, was famously placed 'above the senior wrangler' in the Cambridge mathematical tripos in 1890. (Women were allowed to sit examinations at Cambridge University at the time, but were not classed or awarded degrees: the senior wrangler was the man with top marks.) It's not known when he visited Down: it may have been in 1867. Emma wrote to Henrietta:

My dear Body
What a day yesterday was. I never did. The boys got thro' it wonderfully & in the afternoon Sir John [*Lubbock*] & Prof. Fawcett called & asked G. [*George*] to dinner (& you too) He is the man we saw at the Pop.[*Popular concert of classical music*] that day, very animated & agreeable. He said if he lived 20 or 30 years longer he fully expected that women w^d be regularly taking their degrees at Ox. & Cam. like men (not living in the Colleges of course) He spoke of the female exams as most useful, especially for governesses, & said that experience contradicted the expectations formed viz. that the girls w^d try for a smattering in many lines, for they generally kept to one & did it more thoroughly than men. ... He says he always likes to see women at his lectures & that many of them are attended by women

The civil engineer Edward Cresy had written to Emma about Elizabeth Garrett in 1865.

Metropolitan Board of Works | Spring Gardens
20 Nov 65

My dear Mrs Darwin,
Permit me to awaken your feminine sympathies in behalf of a very admirable young lady & dear friend of ours Miss Elizabeth Garrett who has, after encountering an amount of opposition which few men would have had the courage to encounter, succeeded in obtaining the diploma of the Apothecaries Company, & has started in practice—
Your brother in law Mr Erasmus Darwin is Chairman of the Council of the Bedford College for Girls & Miss Garrett is a candidate for their Professorship of physiology applications for which are to go in on Wednesday next— I have no doubt that if a properly qualified lady can be obtained that the council would be disposed to consider her as possessing many advantages for instructing girls, especially in that particular branch— I cannot of course ask you to urge the claims of one who is a stranger to you—but if you could say

to Mr Erasmus Darwin that you believe that my recommendation would not be lightly given and that I have had the opportunity of watching Miss Garretts career closely & of frequently observing & testing her scientific acquirements and know them to be of a high order—& also that her industry & zeal are beyond all praise, I think possibly that even at second hand such testimony might do her good service— She is frequently at our house & my wife & I both entertain the greatest regard for her—

I may add what I know will interest you although it cannot help her in the matter now under consideration, viz that the very special career to which she has devoted herself has nothing impaired the charm of her manner or her social converse   she is neither masculine nor pedantic & except you knew her intimately you would only recognise a well bred English Lady— I hope you will be able to give me a more favorable account of Mr Darwin than the last. pray remember us both most kindly to him—

Yours very truly | E Cresy

Henrietta wrote to her cousin Hope Wedgwood about men on 13 September 1867:

I think they are born selfisher & therefore I judge 'em by a diff! standard & thk no worse of them nor of myself—if it is original sin tisn't their fault—& I thk they have some merits we have not— I really believe without public schools universities & other inventions of the Devil they wdn't be found so v. m. worse than we are—as it is they are—but that is their misfortune not their fault— As you've been abusing y^r young men I'm going to do the other thing to mine— I really do think George is no worse than he was before the university—you feel that kind of consciousness that you have with a girl & *not* with a boy, that there are *no* hidden depths of wickedness. I don't feel it with all of 'em but I do with George just precisely as m. as if he was one of us— He is m. the m. good hearted of every one of us—& I know he s many virtues   I might try for a 100 years & I shd never get a truthfulness of nature wh. is not the shallowness of Frankian simplicity—but the truthfulness of clear waters with no muddy bottom to hide—and an utter absence of all malignity & asperity—wh. I take to be a manlike virtue— Well as young men are in such bad odour with you I won't go on with my Hymn of Praise

'Frankian simplicity' is annotated 'Barlastian', which probably indicates that it is a reference to Francis Wedgwood of Barlaston, Hope and Henrietta's uncle: a later history describes him as 'straightforward, brusque and unaffected' (Wedgwood and Wedgwood 1980, p. 242). Henrietta thought university was positively bad for men's characters in general:

possibly because the absence of home influences gave more opportunity for vice. With such views it was unlikely that she would have wanted a standard university education herself. Fawcett too, although he was in favour of women's education, seems to have thought at one stage that living away from home was problematic for women (later he was involved with the establishment of Newnham College, Cambridge, a women's college).

This letter from Elinor Bonham Carter in 1870 throws another light on the discussions Henrietta was having with her friends. Elinor and her future husband, Albert Venn Dicey, were also involved in the setting up of Newnham. The sense of the discussion is sometimes hard to grasp but the references to 'lodging life' suggest that living arrangements are at issue, and that Henrietta's experiences as a traveller have shed new light on the problem. Later Elinor goes on to discuss a meeting with a deeply religious woman, the Quaker Caroline Stephen, who was later Virginia Woolf's aunt; and a mathematical approach to marriage. Effie, also mentioned in the letter, was Hope Wedgwood's sister. Snow Wedgwood, another sister, had been teaching classics briefly at the women's college at Benslow House, Hitchin, that later became Girton College, Cambridge (Hirsch 1998, p. 256).

Ravensbourne
March 27/70

Dearest H.

Your letter was very acceptable though as I said before I find it decidedly trying to read yr descriptions. Our winter is never, never ending & one does nothing but have colds & get to feel more & more torpid. We have had deep snow twice lately & east wind in between. But I won't be too gloomy all things come to an end & I feel as if once in my life I must go abroad for a winter, if you are going to take to it as a habit. You won't be able to keep me off joining you some time. I am so glad you are comfortable & like yr independent life. I have come to think less well of real lodging life than I did once. I'm sure many a man feels it very dreary & it is greatly that life that makes 'em so selfish. Oh why am I writing such trite things—what I mean is, I think a modified lodging life is what one wants for men & women, then yours now is as you say modified. …

The next thing I want to talk to you abt is a nice little visit I paid last week to Caroline Stephen (Effie's friend). You know she came here last year & so I got my return invite & staid 3 days with them in London. I was immensely interested & taken with her—& she seemed to have so much to say to me, all at once without any preliminaries, wh. is so nice. I gave myself up entirely to talking with her—& did very little else but a Saturday concert, which by the bye I enjoyed mightily—in spite of a fearful struggle for places. Jo [*Joseph Joachim*] & [*Clara*] Schumann, Mendel[*ssohn*] quartett strings—solos Schumann & Kreuzer Sonata— How curious is the effect that a very tall large person has at first in making one think them ever so much older wiser & better than oneself, at least so it is with me   I didn't mean to say that Caroline S is not all this, but I cant help taking it for granted at first, without knowing it, & wondering what such a superior being can want with a little thing like me! This feeling wore off very soon— I suppose you know she is a regular, conservative woman & will have "none of these things" such as you & I wish to see. education is her special abhorrence, but I don't mind that the least in the world, in fact it is rather an additional interest to me to talk to a woman with quite opposite views— We talked immensely ab^t marriage too & there again her feelings are essentially womanly, of course we came to no more conclusions than you & I do but she hit upon rather a neat way of defining the 1st & second rate marriages, that we so often discuss, by numbers eg. If you feel you are capable of a marriage that is of affection = 50, ought you ever to accept one which on your side is only = 15. She says she thinks she would, for if the man were at 50 in his affections the strong wish on his side would be sufficient to satisfy her in fact to make it impossible for her to refuse. I daresay this view is true of a great many people & I do think the answer to the question I put above depends so very much upon whether the woman is one who has felt all her life, that marriage *is the one* thing she wishes for (in the best sense) if this is the case I think she had better take 15 on the whole rather than wait all her life for 50, just as in everything in this world we have to content ourselves with lowering our ideal to what we can get. Miss Thackeray [*Anne Isabella Thackeray*] came & spent a quiet evening, which was extremely entertaining   I liked her of course very much, who does not? & she being so intimate with C.S the talk was easy directly—She said she had not enough spinster friends! who were "boon companions" & we enrolled ourselves in a society of this name. I heard great deal abt C.Ss book [*The service of the poor (1871), a study of religious sisterhoods*] which interested me very much. she wont get it out just yet though. Also a great deal ab^t her admiration & liking for Effie & more distant admiration of Hope, because she did not know her as she did Effie. I hear Snow is come back from Hitchin to finish up her book [*a biography of John Wesley*], she seems a good deal puzzled in her views ab^t College life, we must try & get her here to hear all abt it. Do you know that they are setting up at Cambridge too after all. I dont mean that Miss Davies [*Emily Davies*] is doing it, but the Professors have instituted regular Lectures for Women & now there is to be a lady there who will receive students. [*Henry Sidgwick rented a house at 74 Regent Street, Cambridge, where women who came from a distance to attend his 'ladies' lectures' could live.*] I wonder how it will answer, it would be just the thing for such as you or me who might want to study & not to be hampered or examined but whether the public in general are prepared for such a move I dont know.

   …

Dont be very long afore you write to yr disconsolate friend & come home in reasonable time—

Yr EMBC

Elizabeth, Darwin's younger daughter, travelled up to London for lectures at the university: she wrote to Horace on 12 February 1873, 'I am going to be so spirited as to go up once a week for a lecture on French history. I come into the middle of a course so I shall only get about six' (DAR 258: 556). On 28 February, she told Horace how the lectures were going:

I was in London on Monday and Tuesday   I stayed two days with Henrietta   and went to my lecture which was interesting but I have such fearful difficulty always to find my way   it is at the University College and it is a complicate way along passages and I have the dread of bursting into some of the men's class rooms which would be awkward. He has a very small audience to lecture to which must make it very flat for him poor man.

It was a fearful day as there had been a heavy snow   I had a cab with two horses. I also went to see some pictures the Old Masters for the last time and a collection of Masons where Uncle Rase's donkey [*The unwilling playmate, a painting by George Mason*] looks better than it ever does in his drawingroom, some of his pictures are very lovely particularly The Harvest Moon which was at the Academy.

The young women in Darwin's circle seemed to have a positive preference for 'difficult' subjects. Elinor Bonham Carter was interested in art, but also took examination courses in geometry. She commented of the lecturer, 'he did put such laudatory remarks on my paper, that I shd become conceited only that I'm sure tis only that he did not think the female mind capable of taking in a Geometrical idea before he gave these Lectures, for there was nothing much in what I did' (DAR 219.8: 13). Francis Darwin's fiancée, Amy Ruck, also seemed to be meditating a course of serious study, as this letter from both of them to Horace Darwin (Jim) of 23 April 1873 shows. Rose and Lucy were Wedgwood cousins.

Pantlludw, | Machynlleth.

An early answer will oblige as we want to confabulate— I go back to Babylon the accursed soon—

Dear Jim

… I wish you would find out a little about Maths   would Stuart be the man who wd teach Amy? It would be very jolly if he would. Rose appears to find it such awfully hard work. could you find out what Stuart thinks about that— Also must she wait till October to begin or can she begin any time; any information about it will be gratefully received   You might find out whether they would teach

Heat & Electric to Amy after she had shewed her brains by math[es]. & if we could get Lucy & Rose to join   Amy doesn't much want mathem unless she can use them as a stepping stone to Physics, if she can't I think she would go in for learning Literature. If you can get a prospectus without bother wd you send one, Amy put away the other so safely she can't find it— If Stuart knows about how much time a day it ought to take a person would you let us know—Amy **says** she works slowly

Yrs | F D

Remember me to Stuart

Dear Horace.

    … I am very sorry that I have mislaid the prospectus—if it is any trouble, never mind ab[t] getting another, as it is pretty sure to turn up, if I hunt.

    Thank you very much for enquiring abt things. | A R. R.

On her travels with her husband in Europe in 1873, Henrietta met Caroline and James Stansfeld, and Josephine Butler. The Stansfelds had edited their marriage ceremony to reflect their view that marriage was a union of equals. James was a leading male activist in the women's rights movement; Caroline was a member of many campaigning organisations. Together with Josephine Butler, they were campaigning against the Contagious Diseases Act, which allowed the compulsory examination of women suspected of being prostitutes (effectively, any women outdoors after dusk) in a number of ports and garrison towns. Caroline Stansfeld shared her cigarettes with Henrietta, which was very daring. (Caroline's father took a progressive view of women's rights and allowed his daughters to smoke cigars.) 'I & M[rs] Stansfield smoked in the most brazen way. They would leave me ½ dozen which R. says he shall throw away to put temptation out of my way' (DAR 245: 5), Henrietta wrote on 26 August to her mother; on 29 August she continued:

> I don't think I've written for 2 days which have been full of events for Villars, namely the appearance of the great Josephine Butler with a frightfully ugly son & M[r] Stewart [*James Stuart, another campaigner against the Contagious Diseases Act; an assistant tutor at Trinity College, Cambridge, when Francis Darwin was there*] Frank's beloved M[r] Stewart if you remember about him. They came up to see the Stansfields for a night M[r] Stansfield is I gather a regular out & outer, & I think that your opinions about Woman are a shibboleth without wh. you can't enter the inner circle. Not but what she is not very civil to me introduced & made Josephine come & sit by me on the sofa—but then I think she thinks I'm v. young & may be converted. However whatever were their opinions they made a very pleasant party. There is somethg particularly taking in M[r] Stewart,

such a fresh bright nature, not at all like an ordinary bored yng man, & very young looking. R. was v. glad to meet him as he had heard he was open to nobbling for W.M.C. [*Working Men's College*] & so they had a good talk. We first dined together & then went into the Stansfields salon & staid ever so late. Josephine has the remains of great beauty & I can quite understand how she captivates mens souls. There is something very commanding & fascinating in her eyes. She told me a little about that wonderful Theodosia Marshall who talked to me about squashing fools at the W.M.C.. She must be a very remarkable woman—tho' she is the dau^r of an employer, & an employer of the old school. She has got the complete confidence of the men & they tell her everything—about their trades union & everything else & talk of her as Theodawsia Marshall. I am sure Horace's Marshall can't be sound in his views about women. She talked about him with just the slightly disparaging tone that Theodosia did.

Theodosia Marshall, otherwise little known, seems to have been involved with women's higher education in Leeds. Her brother, William Cecil Marshall (Horace's Marshall of the unsound views), was an architect, and built a new billiard room onto Down House in 1877. In 1874, she sent Darwin observations on and specimens of insectivorous plants; he used some of her information in *Insectivorous plants*.

There even seems to have been a plan to get Henrietta involved in the Working Women's College founded in London in 1874, in association with her husband Richard's Working Men's College: unfortunately nothing is known of her involvement other than this brief note from Emma:

> About the Women's College F.[*father*] thinks & so do I that if u were a strong woman it w^d be a real good thing to do; but if it ended in your being the real manager in the same way that R is for his college, it w^d be too gt a weight for your little back, & undertaking a thing & giving it up, I suppose, w^d really do more harm than good.

Although no campaigner, Emma certainly didn't view the women's suffrage movement with the abhorrence that some more conservative women did. This note to Henrietta, written on 30 May 1876, exemplifies her practical and rather tranquil attitude.

> Snow has been both agreeable & entertaining. Yesterday we read aloud her pamphlet on suffrage, which a good deal converted W^m & me— Thinking as I do that the game laws & land monopoly is one of the greatest drawbacks to happiness in the country, & as women in general will have no sympathy with either, I think suffrage will be an additional handle against them—

Darwin's own views on such pragmatic support for women's suffrage would have been slightly different. Although both he and Emma were fundamentally opposed to cruelty to animals, Darwin was alarmed at the popular success of the anti-vivisection movement, led by campaigners such as Frances Power Cobbe. He was concerned that a outright ban on vivisection would halt progress in physiological research, and thought that women were ill informed about the benefits of scientific research. He wrote the following (unpublished) letter to *The Times* at the height of the debate. It was typed on an early typewriter that only used capital letters.

TO THE EDITOR OF THE TIMES

SIR,

AS EVERY ONE WHO IS CAPABLE OF FORMING A SOUND JUDGMENT ON THE SUBJECT IS CONVINCED THAT THE RELIEF OF HUMAN SUFFERING IN FUTURE DEPENDS CHIEFLY ON THE PROGRESS OF PHYSIOLOGY, I HOPE THAT YOU WILL FIND SPACE IN YOUR COLUMNS FOR THE ENCLOSED ARTICLE BY DR. RICHARDSON, WHICH HAS JUST APPEARED IN "*NATURE*". THE ARTICLE SHEWS IN A PRACTICAL MANNER, AND MORE CONCLUSIVELY THAN ANYTHING THAT HAS BEEN PUBLISHED ELSEWHERE, THE NECESSITY OF EXPERIMENTS ON LIVING ANIMALS. WOMEN, WHO FROM THE TENDERNESS OF THEIR HEARTS AND FROM THEIR PROFOUND IGNORANCE ARE THE MOST VEHEMENT OPPONENTS OF ALL SUCH EXPERIMENTS, WILL I HOPE PAUSE WHEN THEY LEARN THAT A FEW SUCH EXPERIMENTS PERFORMED UNDER THE INFLUENCE OF ANAESTHETICS, HAVE SAVED AND WILL SAVE THROUGH ALL FUTURE TIME, THOUSANDS OF WOMEN FROM A DREADFUL AND LINGERING DEATH. IT IS HUMILIATING TO REFLECT THAT THOSE TO WHOM MANKIND OWE THE DEEPEST DEBT OF GRATITUDE SHOULD NOW BE OVERWHELMED BY FALSEHOOD AND CALUMNY

I AM, SIR, | YOUR OBEDIENT SERVANT | CHARLES DARWIN

DOWN, BECKENHAM

JUNE 23RD. 1876.

In 1877, a Mrs D, possibly Elinor Dicey (formerly Elinor Bonham Carter) wrote to Darwin about the advisability of women's observing vivisection; the query may have been to do with Newnham Hall (later Newnham College), Cambridge, which opened to students in 1875. Elinor and her husband were involved in the foundation of the college. Mrs D's letters are lost, but Darwin's draft reply survives.

My dear M*rs* D.

In answer to your two letters to my wife I have pleasure in giving you my opinion, which you can communicate to anyone, as you may think fit; but I must beg you to observe that I am not a physiologist & that my opinion can have no special value.—

I should regret that any girl who wished to learn physiology sh*d* be checked, because it seems to me that this science is the best or sole

Elinor Dicey. (Chrystal Album no. 1 PH/10/3.)
By permission of The Principal and Fellows, Newnham College, Cambridge.

one for giving to any person an intelligent view of living beings, &
thus to check that credulity on various points which is so common
with ordinary men & women.

I sh^d look at it as a Sin to discourage any boy from studying physiology who had the wish to do so; & I make the distinction between
a boy & a girl, because as yet no woman has advanced the Science.
I believe much physiology c^d. be learned without seeing any experiments performed or any organ in action; but I do not believe that a
person could learn several parts of the subject with ⟨the⟩ vividness &
clearness, which is necessary for *well* instructing others, unless he saw
some of the organs in action.— All that I have said here with respect
to ordinary students applies with greatly increased force to medical
students; though no doubt very many perhaps most medical men
practice their profession by the mere rule of thumb. With respect
to you not liking a girl to see an animal operated on, though quite
insensible, I can quite understand it & sh^d. sympathise fully with you,
if it were out of mere idle curiosity; but if a person with a wish ⟨to⟩
learn physiology was thus prevented, I sh^d. consider it a weakness.—
I may add that I have bitterly repented this very weakness in my own
case, as I c^d. not get over my horror at seeing men dissected when I
was young.— Even to take the extreme case of an animal becoming
sensible before the operation was over, it w^d take only a few seconds
either to kill it or render it again insensible. Nor can I see the least
reason to suppose it w^d suffer more during such few seconds than it
w^d for hours during any severe illness to which men & animals are

liable. By dwelling too much on humanity, though Heaven knows this until lately has been a rare error, do you not think that there is danger of compassion becoming morbid?

Pray believe me, dear Mrs D | Yours very sincerely | Ch Darwin

Darwin's readers also occasionally asked for his views. Charlotte Papé was probably German; she was in Britain in 1875, and later wrote in German on the lack of mother's rights over their children. In Germany, as in England, women had no right to keep their children after the breaking up of a marriage. She corresponded with Helen Taylor, the stepdaughter of John Stuart Mill. In this letter to Darwin, she voiced a concern that must have occurred to many women: looking around them at their male and female friends and acquaintances, it did not seem obvious to them that the men were the cleverer. Such observations might be criticised as anecdotal, but could they not be set on firmer ground? Unfortunately Papé's paper and Darwin's reply are lost.

Lark Hill House | Edgeley | Stockport
July 16th. 75

Dear Mr. Darwin,

I must ask your pardon and your indulgence for the great liberty I am going to take just now in begging of you the favour to look at the enclosed paper. The general interest which I always took in questions relating to the laws that regulate the developement of life has been raised to a very strong wish to know as much about it as I can, by your own works, which I have only now been able to *really* read; and Mr. Francis Galton's books have shown me, in what, on the whole, simple way facts bearing on some questions of Heredity may be collected. I have been thinking that perhaps even I might be able, by accurately tabulating and comparing such cases as I know, to do something towards ascertaining the truth or error of some of Mr. Galton's conclusions, at least as far as my own conviction is concerned. The point which naturally has the greatest interest for me, about which I am most anxious to find out something certain, is, how far heredity is limited by sex in the human race, especially whether mental qualities are at all limited by it. I am well aware that your own, I think, provisional view is, that even mental qualities are thus limited; I myself know so comparatively many striking instances to the contrary, among my friends and my own family, that it seems highly improbable to me. At any rate, every woman ought to try to ascertain as much of the truth in respect to it as she can; for apart from the interest of the question in itself, it is most important for the future of women.

Now I have noted down different rubrics, as on the paper enclosed, to be filled out as accurately as possible; and the great, very great, favour I am begging of you, dear Mr. Darwin, is just to throw a look at it and tell me, whether, if I do so, the conclusions appearing from such tables would be trustworthy as far as they go; also what number of families would be the minimum for a reliable average, and any other remark necessary, and so invaluable from you, and for me.

For, of course, like all women, I have had no scientific training, and know nothing except from random reading; neither could I attain any now. And it is just this very helplessness as to getting information, or even any word of advice and criticism that I could trust more than my own that must form my excuse for the unwarrantable liberty I am taking, and plead with your kindness for the granting of the favour I beg. I literally know of no one to ask, except the illustrious authority I am addressing; and so doing I wonder at my own boldness. If you think such tables no good, at least if not put together by more skilful hands, of course, I shall not attempt to fill them out.

I am, dear Mr. Darwin, | with true admiration and reverence | Yours | Charlotte Papé

Caroline Kennard was the wife of a businessman in Boston, Massachusetts; she was a member of the New England Woman's Club and the Women's Educational and Industrial Union and participated in meetings of the Association for the Advancement of Women. At one of these meetings, she read a paper, 'Housekeeping a professsion', which argued that housekeeping should be measured in economic terms like any other profession. Her courteous yet forceful engagement with Darwin makes a fitting close to this book.

Mr. Darwin
Dear Sir.

I a paper recently read before a company of women in Boston, ground was taken of the inferiority of women; past, present and future; based upon scientific principles: as concisely reported in the newspaper extract enclosed.

In reply to opposing arguments in the discussion following the paper, the Author stated her scientific Authority to be Mr. Darwin, in his "Origin of Species".

As a believer in continued scientific discoveries and revelations answering and modifying, ultimately, all material questions; and as an admirer of your cautious and candid methods of conveying great results of learning and investigations to the world, I take the liberty to inquire whether the Author of the paper rightly inferred her arguments from your work: or if so, whether you are of the same mind now, as to possibilities for women, judging from her organization &c

If a mistake has been made the great weight of your opinion and authority should be righted: to which, I take it for granted, you would not object.

Excuse the liberty I take of addressing you and the hope of a reply in enclosed envelope.

I am yours with expressions of great esteem | Caroline A. Kennard.

Brookline | Dec. 26. 1881.

Down, | Beckenham, Kent. | (Railway Station | Orpington. S.E.R.)
Jan. 9th. 1882

Dear Madam

The question to which you refer is a very difficult one. I have discussed it briefly in my "Descent of Man". I certainly think that women though generally superior to men to moral qualities are inferior intellectually; & there seems to me to be a great difficulty from the laws of inheritance, (*if I understand these laws rightly*) in their becoming the intellectual equals of man. On the other hand there is some reason to believe that aboriginally (& to the present day in the case of Savages) men & women were equal in this respect, & this w^d. greatly favour their recovering this equality. But to do this, as I believe, women must become as regular "bread-winners" as are men; & we may suspect that the early education of our children, not to mention the happiness of our homes, would in this case greatly suffer.

I have written this letter without any care of style, as it is intended only for your private use.—

Dear Madam | Yours faithfully | Ch. Darwin

Brookline Mass. U.S.A
Jan. 28– 1882–

Mr. Darwin

Dear Sir,

I thank you for your very kind reply to my letter of inquiry as to your opinion of the comparative intellectual abilities of the sexes—

I believe you are supported in your ideas of the greater moral qualities of woman— Before quite deciding as to her condition intellectually will you excuse me if I remind you that recent results from efforts for her higher education, in your own country and in this, are very flattering and encouraging: and are opening for women avenues for individual improvement and for the general enlightenment of her sex— and therefore, of necessity (according to the laws of heredity) for the advancement of the human race intellectually. Her enlightened intellect, united with her wholesome moral nature, can then with the aid of man (for in nature the male & female must work in sympathy together, you have taught us—)—ordain, in a manner hitherto unthought of or practised upon, for the propagation of the best and the survival of the fittest in the human species.

The laws of heredity have been closely watched in the lower animals, and tendencies toward improvement encouraged and toward deterioration guarded against; while in marriages and the begetting of offspring, the perpetuation of the best physical, intellectual and moral tendencies in the human race have been mostly unheeded and neglected—

In reply to your argument that "women must become as regular 'bread-winners" as are men"; have they not been and are they not largely, bread-winners; though unrecognized generally as such?

Partners in business—share money profits and why should not partners in marriage—where the wife, by her labor and economy does her full part toward husbanding for the future? In the unceasing demand upon the head of a household, for executive ability, fixedness of purpose, and courage of execution, are not women possessed of the same kind of qualities which would grow with the using into as *apparent* & grand results as are accorded to men of business, government officials, & army officers and statesmen who all expect compensation for services rendered?

And why be anxious for the "education of our children" and "the happiness of our homes", if women become breadwinners? when in this country five sixths of the educa*tors* are women and acknowledged 'breadwinners', beside improving the condition of their homes and adding happiness thereto—

Which of the partners in a family is the breadwinner where the husband works a certain number of hours in the week and brings home a pittance of his earnings (the rest going for drinks & supply of pipe) to his wife; who, early & late, with no end of self sacrifice in scrimping for her loved ones, toils to make each penny tell for the best economy and besides, to these pennies she may add by labor outside or taken in?

Dr. Walker, once president of Harvard College said that, of the young men who had been by personal effort, assisted through that college, three fourths had been, by efforts of women. And we know it has been the custom for Mothers & sisters to help their sons & brothers, by every possible effort, to an education (Whoever heard of a brother assisting a sister through college while he druged & toiled?

One young woman I know who receives pay for nursing the sick and gives the half of it to a brother who is learning to engrave. Is she less a bread-winner than he—or less than the other brother who, though younger than herself, by aid of the Father & herself received an education which she longed for and that enabled him to rank with our most prominent clergymen?

The family must be *right*eously maintained   Let the 'environment' of women be similar to that of men and with his opportunities, before she be fairly judged, intellectually his inferior, please.—

Excuse this great liberty and I am your obliged | Caroline A Kennard

List of letters and provenances

The following list gives details of the letters featured in the preceding chapters in order of publication, with the name of the correspondent, the date, the location of the original letter, and, where applicable, the volume and page number of the letter as published in the *Correspondence of Charles Darwin*. Dates or parts of dates in square brackets have been supplied by the editors of the Darwin Correspondence Project. At the end of the list is a key to locations.

1. Friends

From Mary Congreve, 27 October [1821], DAR 204: 186 (*Correspondence* 1: 1)

From Fanny Mostyn Owen, [26 September 1831], DAR 204: 52 (*Correspondence* 1: 168)

From Fanny Myddelton Biddulph, 14 January 1837, DAR 204: 57 (*Correspondence* 2: 1)

To Mary Butler, 20 February [1859], Brown University, John Hay Library, A. E. Lownes Manuscript Collection, Ms 84.2, box 3, folder 37 (*Correspondence* 7: 249)

To Mary Butler, 11 September [1859], APS, 168 (*Correspondence* 7: 331)

From Mary Butler, [before 25 December 1862], DAR 160: 392 (*Correspondence* 10: 626)

From H. A. Huxley, 1 January 1868, DAR 166: 284 (*Correspondence* 13: 5)

To H. A. Huxley, 16 October [1872], Imperial College, Huxley 5: 291 (*Correspondence* 20: 447)

From M. C. Stanley (Lady Derby), [16 November 1871], DAR 162: 163 (*Correspondence* vol. 24: Supplement)

To M. C. Stanley (Lady Derby), [18 November 1871], DAR 143: 384 (*Correspondence* vol. 24: Supplement)

From M. C. Stanley (Lady Derby), 4 June 1872, DAR 162: 165 (*Correspondence* 20: 244)

From M. C. Stanley (Lady Derby), 14 September 1875, DAR 162: 167 (*Correspondence* 23: 361)

From M. C. Stanley (Lady Derby), 22 December 1875, DAR 162: 168 (*Correspondence* 23: 502)

From M. C. Stanley (Lady Derby), 19 September 1877, DAR 162: 169 (*Correspondence* vol. 25)

From M. C. Stanley (Lady Derby), 24 May 1878, DAR 162: 170 (*Correspondence* vol. 26)

From Emma Darwin to M. C. Stanley (Lady Derby), 12 November [1879], Liverpool Archives and Family History (920 DER (15) 43/9/23) (*Correspondence* vol. 27)

From M. C. Stanley (Lady Derby), 16 October 1881, DAR 162: 171 (*Correspondence* vol. 29)

From Susan Norton, 20 November [1871], DAR 172: 78 (*Correspondence* 19: 693)

To Susan Norton, 23 November [1871], Houghton Library, Charles Eliot Norton Papers, MS Am 1088 (1594) (*Correspondence* 19: 698)

To Sarah Haliburton, 1 November [1872], DAR 185: 22 (*Correspondence* 20: 475)

From Sarah Haliburton, 3 November [1872], DAR 166: 85 (*Correspondence* 20: 477)

To Sarah Haliburton, 6 November [1872], DAR 185: 23 (*Correspondence* 20: 483)

From Sarah Haliburton, 21 November [1880], DAR 99: 211–12 (*Correspondence* vol. 28)

To Sarah Haliburton, 22 November 1880, DAR 185: 24 (*Correspondence* vol. 28)

From Sarah Haliburton, 12 December [1880], DAR 99: 209–10 (*Correspondence* vol. 28)

To Sarah Haliburton, 13 December 1880, DAR 185: 25 (*Correspondence* vol. 28)

From Sarah Haliburton, 8 September [1881], DAR 166: 87 (*Correspondence* vol. 29)

2. Marriage

From Fanny Mostyn Owen, 2 [December 1831], DAR 204: 54 (*Correspondence* 1: 184)

From Caroline Darwin, 20 December [1831], DAR 204: 70 (*Correspondence* 1: 188)

From Catherine Darwin, 8 and 29 January 1832, DAR 204: 83 (*Correspondence* 1: 192)

From Charlotte Wedgwood, 12 and 29 January 1832, DAR 204: 116 (*Correspondence* 1: 195)

To Caroline Darwin, 2–6 April 1832, DAR 223: 10 (*Correspondence* 1: 220)

From Catherine Darwin, 25 July 1832, DAR 204: 85 (*Correspondence* 1: 254)

From Catherine Darwin, 27 September 1833, DAR 204: 88 (*Correspondence* 1: 334)

To Caroline Darwin, [9 November 1836], DAR 154: 49 (*Correspondence* 1: 518)

From Harriet Henslow, 22 November 1838, DAR 204: 165 (*Correspondence* 2: 124)

From W. E. Darwin to Emma Darwin, 4 March [1869], DAR 210.5: 4

From Emma Darwin to W. E. Darwin, [6 March 1869], DAR 219.1: 85

To H. E. Litchfield, 4 September [1871], British Library, Add 58373 (*Correspondence* 19: 550)

From Emma Darwin to Leonard Darwin, 29 July [1874], DAR 239.23: 1.20

From W. E. Darwin to Emma Darwin, [24 July 1877], DAR 210.5: 16

From Bessy Darwin to Ida Farrer, 12 October [1877], DAR 258: 565

From Emma Darwin to Sara Darwin, [24 January 1878], DAR 219.1: 104

3. Children

To Emma Darwin, [9 May 1842], DAR 210.8: 20 (*Correspondence* 2: 318)

To J. S. Henslow, 17 January [1850], DAR 93: 96–7 (*Correspondence* 4: 303)

From Emma Darwin to H. E. Litchfield, [7 September 1876], DAR 219.9: 139

From Emma Darwin to H. E. Litchfield, [8 September 1876], DAR 219.9: 140

From Emma Darwin to W. E. Darwin, [13 September 1876], DAR 210.6: 144

To Emma Darwin, [1 July 1841], DAR 210.8: 16 (*Correspondence* 2: 293)

From Emma Darwin to F. E. E. Wedgwood, 20 October 1842, *Emma Darwin* (1904) 2: 50

From Catherine Thorley to Emma Darwin, 14 April 1851, DAR 210.13: 5

From Catherine Thorley to Emma Darwin, 16 April 1851, DAR 210.13: 6

From F. E. E. Wedgwood to Emma Darwin, 19 April 1851, DAR 210.13: 15

From Emma Darwin to F. E. E. Wedgwood, 22 April 1851, DAR 210.13: 24

From Emma Darwin to W. E. Darwin, [3 February 1863], DAR 219.1: 70

From Emma Darwin to W. E. Darwin, [*c*. March 1852], DAR 219.1: 2

From Emma Darwin to W. E. Darwin, [24 November 1852], DAR 219.1: 6

From Emma Darwin to G. H. Darwin, 17 August 1884, DAR 210.3: 114

4. Scientific wives and allies

From Emma Wedgwood, [21–2 November 1838], DAR 204: 150 (*Correspondence* 2: 122)

To Emma Wedgwood, [20 January 1839], DAR 210.8: 12 (*Correspondence* 2: 166)

From J. D. Hooker, [2 June 1865], DAR 102: 24–7 (*Correspondence* 13: 172)

From Emma Darwin to W. E. Darwin, [6 September 1879], DAR 219.1: 126

From Emma Darwin to H. E. Litchfield, [19 October 1873], DAR 219.9: 106

From Emma Wedgwood, [30 December 1838], DAR 204: 157 (*Correspondence* 2: 149)

From Marion Bell to F. E. E. Wedgwood, 4 November [1872], DAR 160:

126 (*Correspondence* 20: 517, enclosure to letter from E. A. Darwin, 20 November 1872)

To Emma Darwin, 5 July 1844, NHM, MSS DAR A4 (*Correspondence* 3: 43)

From E. F. Lubbock, [1 October 1866], DAR 170: 8 (*Correspondence* 14: 331)

From E. F. Lubbock, [1867–8?], DAR 104: 227–8 (*Correspondence* 18: 403)

From E. F. Lubbock to Emma Darwin, [*c.* 29 November 1873], DAR 170: 16 (*Correspondence* 21: 523)

From E. F. Lubbock, [after 2 July] 1875, photocopy in the private collection of the Lubbock family (*Correspondence* 23: 246)

From F. H. Hooker, 6 September [1865], DAR 104: 239–40 (*Correspondence* 13: 228)

From F. H. Hooker, 22 September [1865], DAR 104: 237–8 (*Correspondence* 13: 236)

To E. F. Lubbock, [before 7 April 1873], DAR 170: 17 (*Correspondence* 21: 163)

From J. D. Hooker, [7 April 1873], DAR 103: 153–4 (*Correspondence* 21: 165)

From A. B. Buckley, 16 December 1879, DAR 160: 366 (*Correspondence* vol. 27)

To A. B. Buckley, 31 October [1880], DAR 143: 182 (*Correspondence* vol. 28)

From A. B. Buckley, 7 November 1880, DAR 160: 370 (*Correspondence* vol. 28)

From A. B. Buckley, 13 January 1881, DAR 160: 371 (*Correspondence* vol. 29)

5. Observing plants

To D. F. Nevill, 12 November [1861], APS, 270 (*Correspondence* 9: 338)

From D. F. Nevill, [before 22 January 1862], DAR 172: 26 (*Correspondence* 10: 39)

From D. F. Nevill, 2 September [1874], DAR 172: 24 (*Correspondence* 22: 433)

To D. F. Nevill, 7 September 1874, APS, 449 (*Correspondence* 22: 446)

From D. F. Nevill, 8 [September 1874], DAR 172: 20 (*Correspondence* 22: 448)

From D. F. Nevill, 16 [September 1874], DAR 172: 21 (*Correspondence* 22: 461)

To D. F. Nevill, 18 September [1874], Cleveland Health Sciences Library, Robert M. Stecher collection (*Correspondence* 22: 465)

From D. F. Nevill, 22 [September 1874], DAR 172: 22 (*Correspondence* 22: 473)

From L. E. Becker, 18 May 1863, DAR 160: 108 (*Correspondence* 11: 424)

From L. E. Becker, 21 May [1863], DAR 160: 106 (*Correspondence* 11: 425)

From L. E. Becker, 23–4 May [1863], DAR 160: 107 (*Correspondence* 11: 435)

From L. E. Becker, 28 May [1863], DAR 160: 109 (*Correspondence* 11: 457)

From L. E. Becker, 8 July [1863], DAR 160: 110 (*Correspondence* 11: 527)

From L. E. Becker, 14 October 1869, DAR 160: 117, 119 (*Correspondence* 17: 429)

From Mary Treat, 13 December 1872, DAR 58.1: 23–4 (*Correspondence* 20: 570)

To Mary Treat, 1 January 1873, Amy Nagashima, private collection (*Correspondence* 21: 19)

From Mary Treat, 28 July 1873, DAR 58.1: 30–2 (*Correspondence* 21: 303)

To Mary Treat, 12 August 1873, Vineland Historical and Antiquarian

Society (*Correspondence* 21: 325)

From Mary Treat, 8 June 1874, DAR 58.1: 58–9 (*Correspondence* 22: 280)

To Mary Treat, 22 June 1874, Amy Nagashima, private collection (*Correspondence* 22: 306)

From Mary Treat, 2 December 1874, DAR 58.1: 109–10 (*Correspondence* 22: 553)

From Mary Treat, 3 April 1876, DAR 178: 178 (*Correspondence* vol. 24)

To Mary Treat, 21 April [1876], Vineland Historical and Antiquarian Society (*Correspondence* vol. 24)

From Mary Treat, 15 May 1876, DAR 178: 179 (*Correspondence* vol. 24)

To Mary Treat, 1 June 1876, Amy Nagashima, private collection (*Correspondence* vol. 24)

From S. B. Herrick, 12 February 1876, DAR 166: 189 (*Correspondence* vol. 24)

To S. B. Herrick, 6 March 1876, University of Virginia Library, Special Collections, MS dept, 3461 (*Correspondence* vol. 24)

6. Companion animals

From J. L. Gray, 14 February 1870, DAR 80: 162–3 (*Correspondence* 18: 34)

To Lucy Wedgwood, 8 June [1867–72], CUL, Add 4251: 334 (*Correspondence* 20: 601)

From A. J. Cupples to Emma Darwin, 8 November [1872], DAR 161: 281 (*Correspondence* 20: 488)

From C. J. Shuttleworth, 27 November [1871–80], DAR 177: 158 (*Correspondence* vol. 30, Supplement)

To F. P. Cobbe, 28 November 1872, The Huntington Library, CB 386 (*Correspondence* 20: 528)

From F. P. Cobbe, 28 November 1872, DAR 161: 187 (*Correspondence* 20: 529)

From Athénaïs Michelet, 17 May 1872, DAR 171: 170 (French) (*Correspondence* 20: 212)

To Athénaïs Michelet, 23 May 1872, APS, 417 (*Correspondence* 20: 224)

From Athénaïs Michelet, 26 June 1872, DAR 171: 171 (French) (*Correspondence* 20: 282)

From Dora Roberts, 17 December [1872 or later], DAR 176: 184 (*Correspondence* 20: 576)

From Pauline Perfilieff (Praskov'ja Perfil'eva), 22 February 1874, DAR 174: 36 (French) (*Correspondence* 22: 105)

From Thereza Story-Maskelyne, 4 May 1874, DAR 177: 263 (*Correspondence* 22: 238)

From G. A. Wolfe, 9 March 1875, DAR 181: 135 (*Correspondence* 23: 97)

7. Insects and angels

To M. E. Lyell, [4 October 1847], APS, 63 (*Correspondence* 4: 81)

To M. E. Lyell, [24 October 1849], DAR 146: 332 (*Correspondence* 4: 272)

To M. A. T. Whitby, 2 September [1847], APS, 61 (*Correspondence* 4: 63)

To M. A. T. Whitby, 14 October [1847], Lehigh (*Correspondence* 4: 89)

To M. A. T. Whitby, 12 August [1849], NYAM, MS 15 (*Correspondence* 4: 248)

From M. H. Morris to R. C. Alexander, 17 June 1855, DAR 205.2: 247 (*Correspondence* 5: 355)

From Asa Gray, 30 June 1855, DAR 165: 92a (*Correspondence* 5: 362)

To K. M. Lyell, 26 January [1856], APS, 124 (*Correspondence* 6: 32)

To the Linnean Society, 10 May 1869, Linnean Society of London Archives (*Correspondence* 17: 224)

From Mary Treat, 20 December 1871, DAR 58.1: 33 (*Correspondence* 19: 727)

To Mary Treat, 5 January 1872, Amy Nagashima, private collection (*Correspondence* 20: 11)

From Mary Treat, 13 December 1872, DAR 58.1: 23–4 (*Correspondence* 20: 570)

From L. C. Wedgwood, 20 November [1871], DAR 181: 62 (*Correspondence* 19: 694)

To L. C. Wedgwood, 5 January [1872], CUL, Add 4251: 331 (*Correspondence* 20: 12)

From L. C. Wedgwood, 20 January [1872], DAR 94: 1b (*Correspondence* 20: 32)

To L. C. Wedgwood, 21 January [1872], CUL, Add 4251: 332 (*Correspondence* 20: 34)

From L. C. Wedgwood, [15 June 1872?], DAR 181: 61 (*Correspondence* 20: 258)

To Sophy Wedgwood, 8 October [1880], CUL, Add 4251: 335 (*Correspondence* vol. 28)

From Sophy Wedgwood, 15 October [1880], DAR 181: 69 (*Correspondence* vol. 28)

From M. W. Tanner, 12 December 1881, DAR 178: 51 (*Correspondence* vol. 29)

From Ada Kepley, 20 February 1882, DAR 169: 7 (*Correspondence* vol. 30)

8. Observing humans

From Snow (F. J.) Wedgwood to H. E. Darwin, [1867–72], DAR 189: 140 (*Correspondence* 20: 597)

From Snow (F. J.) Wedgwood to H. E. Darwin, [1867–72], DAR 181: 45 (*Correspondence* 20: 598)

From Anne Barnard, 30 March 1871, DAR 160: 42 (*Correspondence* 19: 237)

From L. M. Forster to H. E. Litchfield, 20 February 1873, DAR 164: 159 (*Correspondence* 21: 79)

From J. L. Gray, 9 May 1869, DAR 165: 167–8 (*Correspondence* 17: 218)

From M. I. Snow, 29 [November 1872 or later], DAR 177: 213 (*Correspondence* 20: 533)

From L. F. Kempson to Emma Darwin, 20 June 1867, DAR 169: 4 (*Correspondence* 15: 306)

From C. M. Hawkshaw to Emma Darwin, 9 February 1868, DAR 166: 121

(*Correspondence* 16: 88)

From E. M. Bonham-Carter, 20 March [1868], DAR 160: 244 (*Correspondence* 16: 284)

From C. M. Hawkshaw to Emma Darwin, 12 April 1868, DAR 166: 122 (*Correspondence* 16: 393)

From M. S. Vaughan Williams to H. E. Darwin, [after 14 October 1869], DAR 180: 4 (*Correspondence* 17: 433)

From L. C. Wedgwood, [January 1871], DAR 181: 57 (*Correspondence* 19: 8)

From A. M. Lane Fox to E. F. Lubbock, 25 July [1875], DAR 164: 170 (*Correspondence* 23: 295)

From A. M. Lane Fox, 3 August 1875, DAR 164: 171 (*Correspondence* 23: 310)

To Annie Dowie, 1 August 1875, National Library of Australia, MS 760/2/10–11 (*Correspondence* 23: 307)

From Annie Dowie, 10 August [1875], DAR 162: 240 (*Correspondence* 23: 315)

To Emily Talbot, 19 July 1881, University of Chicago Library (*Correspondence* vol. 29)

9. Editors

From Caroline Darwin, 28 October [1833], DAR 204: 78 (*Correspondence* 1: 345)

From Susan Darwin, 12 February 1834, DAR 204: 102 (*Correspondence* 1: 366)

To John Murray, 5 April [1859], Murray Archive, Ms.42152 ff.35–35A (*Correspondence* 7: 278)

To J. D. Hooker, 11 May [1859], DAR 115.1: 15 (*Correspondence* 7: 296)

To A. R. Wallace, 12 July [1871], British Library, Add 46434 (*Correspondence* 19: 484)

To H. E. Darwin, 26 July [1867], DAR 185: 57 (*Correspondence* 15: 329)

To H. E. Darwin, [8 February 1870], British Library, Add 58373 ff.1–2 (*Correspondence* 18: 25)

From H. E. Darwin, [after 8 February 1870], DAR 245: 33b (*Correspondence* 18: 25)

To H. E. Darwin, [March?] 1870, DAR 185: 58 (*Correspondence* 18: 55)

To H. E. Darwin, 20 March 1871, DAR 153: 77 (*Correspondence* 19: 199)

From H. E. Darwin, 21 March [1871], Cornford family papers, private collection (*Correspondence* vol. 24, Supplement)

To H. E. Litchfield, 25 July 1872, John Wilson, dealer (*Correspondence* 20: 322)

To H. E. Litchfield, 21 [March 1874], DAR 153: 84 (*Correspondence* 22: 165)

From R. M. Kettle, 10 March [1871], DAR 169: 8 (*Correspondence* 19: 156)

From Lawson Tait, 5 June [1875], DAR 178: 8 (*Correspondence* 23: 222)

10. Writers and critics

From Eliza Meteyard, 25 April 1865, DAR 171: 160 (*Correspondence* 13: 130)

From Eliza Meteyard, 17 November 1865, DAR 171: 161 (*Correspondence* 13:

309)

From Eliza Meteyard, 19 February 1869, DAR 171: 162 (*Correspondence* 17: 87)

From Eliza Meteyard, 20 April 1874, DAR 171: 163 (*Correspondence* 22: 216)

From Eliza Meteyard, 27 June 1874, DAR 171: 164 (*Correspondence* 22: 332)

From L. E. Becker, 30 March 1864, DAR 160: 112 (*Correspondence* 12: 103)

To M. E. Lyell, [19? October 1866], Bodleian, Dep. c.370, folder MSD-1 (on loan from Somerville College) (*Correspondence* 14: 355)

From Mary Somerville, 30 October 1866, DAR 177: 217 (*Correspondence* 14: 365)

To Mary Somerville, 21 January [1869], Bodleian, Dep. c.370, folder MSD-1 (on loan from Somerville College) (*Correspondence* 17: 33)

From F. P. Cobbe, 28 March [1870?], DAR 161: 186 (*Correspondence* 18: 84)

From E. J. Pfeiffer, [before 26 April 1871], DAR 174: 40 (*Correspondence* 19: 336)

To E. J. Pfeiffer, 26 April [1871], Field Museum, Misc 14 (*Correspondence* 19: 338)

From F. J. Wedgwood to H. E. Darwin, 1 April 1871, DAR 88: 68–70 (*Correspondence* 19: 246)

From Emma Darwin to H. E. Litchfield, [4 December 1873], DAR 219.9: 109

To Marian Evans (George Eliot), 30 March [1873], Armacost Library (*Correspondence* 21: 144)

From Marian Evans (George Eliot), 21 March [1873], Yale, Beinecke, George Eliot and George Henry Lewes Collection, GEN MSS 963, box 2 (*Correspondence* 21: 146)

To A. B. Buckley, 11 February [1876], DAR 143: 179 (*Correspondence* vol. 24)

From A. B. Buckley, 12 February 1876, DAR 160: 365 (*Correspondence* vol. 24)

To A. B. Buckley, 14 November 1880, DAR 143: 184 (*Correspondence* vol. 28)

11. Religion

From Caroline Darwin, [22 March 1826], DAR 204: 20 (*Correspondence* 1: 35)

To Caroline Darwin, 8 April [1826], DAR 154: 29 (*Correspondence* 1: 39)

From Emma Wedgwood, [23 January 1839], DAR 204: 162 (*Correspondence* 2: 169)

From Emma Darwin, [*c.* February 1839], DAR 210.8: 14 (*Correspondence* 2: 171)

From Emma Darwin, [June 1861], DAR 210.8: 35 (*Correspondence* 9: 155)

From Emma Darwin to W. E. Darwin, [6 May 1859], DAR 210.6: 44

From Emma Darwin to H. E. Darwin, [26 September 1866], DAR 219.9: 47

Elizabeth Darwin to Horace Darwin, [winter 1869–70?], DAR 258: 563

Emma Darwin to G. S. Ffinden, [22? November 1873], Bromley, P/123/25/3/1/1 (*Correspondence* 21: 517)

Emma Darwin to Horace Darwin, [29 November 1873], DAR 258: 585

From Emma Darwin to J. B. Innes, 12 October [1874], Cleveland Health

Sciences Library, Robert M. Stecher collection (*Correspondence* 22: 494)

From Emma Darwin to J. B. Innes, 24 December [1875], Cleveland Health Sciences Library, Robert M. Stecher collection (*Correspondence* 23: 504)

From M. E. Boole, 13 December 1866, DAR 160: 249 (*Correspondence* 14: 423)

To M. E. Boole, 14 December 1866, DAR 143: 121 (*Correspondence* 14: 425)

From M. E. Boole, 17 December [1866], DAR 160: 250 (*Correspondence* 14: 432)

From Mary Jung, 7 January 1879, DAR 168: 94 (*Correspondence* vol. 27)

To Mary Jung, 11 January 1879, Stargardt (dealer), catalogue 681 (28 and 29 June 2005), lot 401 (*Correspondence* vol. 27)

From Emma Darwin to H. E. Darwin, [8 February 1870], DAR 219.9: 72

From Emma Darwin to H. E. Darwin, 21 February [1870], DAR 219.9: 74

From Emma Darwin to F. P. Cobbe, [25 February 1871], The Huntington Library, Frances Cobbe Collection CB 390 (*Correspondence* 19: 106)

From H. E. Darwin, [13 November 1871], Cornford family papers, private collection (*Correspondence* vol. 24: Supplement)

From Emma Darwin to W. E. Darwin, 20 January 1885, DAR 219.1: 178

From Emma Darwin to W. E. Darwin, [20 January 1885], DAR 219.1: 177

From Emma Darwin to W. E. Darwin, [25 January 1885], DAR 219.1: 179

From Emma Darwin to W. E. Darwin, [26? January 1885], DAR 219.1: 180

12. Travel

From H. E. Darwin to Emma Darwin, 4 June [1866], DAR 219.10: 18

From M. B. Bathoe, 25 March [1871], DAR 87: 31–6 (*Correspondence* 19: 212)

From F. C. Dixie, 29 October 1880, DAR 162: 182 (*Correspondence* vol. 28)

From F. C. Dixie, 4 November [1880], DAR 162: 183 (*Correspondence* vol. 28)

From F. C. Dixie, 29 November [1880], DAR 162: 184 (*Correspondence* vol. 28)

To Marianne North, 2 August 1881, North 1893, 2: 216 (*Correspondence* vol. 29)

13. Servants and governesses

From Jessie Brodie to H. E. Litchfield, 22 November 1871, DAR 219.8: 27

From F. E. E. Wedgwood, 25 April [1851], DAR 210.13: 33 (*Correspondence* 5: 28)

To E. M. Thorley, 26 April [1851], DAR 148: 73 (*Correspondence* 5: 29)

From Emma Darwin to W. E. Darwin, [23 February 1859], DAR 210.6: 36

From H. E. Darwin to W. E. Darwin, [after 14 March 1859], DAR 210.6: 41

From Emma Darwin to W. E. Darwin, [24 March 1859], DAR 210.6: 42

From Emma Darwin to W. E. Darwin, [24 or 31 October 1859], DAR 210.6: 50

From Emily Thorley to H. E. Darwin, 1863, DAR 219.8: 2

To Camilla Ludwig, 21 February [1862 or later], APS, 620 (*Correspondence* vol. 30, Supplement)

To Camilla Ludwig, 22 December 1862, R. M. Smythe, dealer (*Correspondence* 13: 459)

From Emma Darwin to W. E. Darwin, [27 May 1862], DAR 219.1: 57

To Camilla Ludwig, 26 August [1862], APS, B/D25.272 (*Correspondence* 10: 384)

From Emma Darwin to W. E. Darwin, [6 November 1862], DAR 219.1: 64

From Emma Darwin to W. E. Darwin, [19 November 1862], DAR 219.1: 68

From Emma Darwin to W. E. Darwin, [14 September 1883], DAR 219.1: 160

From J. D. Hooker, [5 May 1862], DAR 101: 33, 134a (*Correspondence* 10: 180)

From J. D. Hooker, 9 June 1862, DAR 101: 40–1 (*Correspondence* 10: 237)

From J. D. Hooker, 19 [June 1862], DAR 101: 38–9 (*Correspondence* 10: 258)

From J. D. Hooker, 28 June 1862, DAR 101: 42–3 (*Correspondence* 10: 275)

From J. D. Hooker, 2 July 1862, DAR 101: 44–5 (*Correspondence* 10: 294)

From J. D. Hooker, [15 and] 20 November [1862], DAR 101: 79 (*Correspondence* 10: 527)

From J. D. Hooker, 18 October 1867, DAR 102: 180–1 (*Correspondence* 15: 399)

To J. B. Innes, 10 December [1868], Cleveland Health Sciences Library (Robert M. Stecher Collection) (*Correspondence* 16: 888)

From Emma Darwin to V. L. Isitt, [before 17 September 1871], DAR 96: 101 (*Correspondence* 19: 579)

From Emma Darwin to H. E. Litchfield, [17 September 1871], DAR 219.9: 95

From H. E. Litchfield to Leonard Darwin, 11 August 1874, DAR 258: 1640

14. Ascent of woman

From L. E. Becker, 22 December 1866, DAR 160: 113 (*Correspondence* 14: 435)

From L. E. Becker, 6 February 1867, DAR 160: 115 (*Correspondence* 15: 68)

From Emma Darwin to H. E. Darwin, [*c.* 20 April 1867], DAR 219.9: 53

From Edward Cresy to Emma Darwin, 20 November 1865, DAR 161: 247 (*Correspondence* 13: 309)

From H. E. Darwin to H. E. Wedgwood, 13 September 1867, DAR 219.10: 22

From E. M. Bonham Carter to H. E. Darwin, 27 March 1870, DAR 219.8: 21

From Elizabeth Darwin to Horace Darwin, 28 February 1873, DAR 258: 557

From Francis and Amy Darwin to Horace Darwin, [23 April 1873], DAR 258: 790

From H. E. Litchfield to Emma Darwin, [29 August 1873], DAR 251: 2337

From Emma Darwin to H. E. Litchfield, [25 April 1874], DAR 219.9: 111

From Emma Darwin to H. E. Litchfield, [30 May 1876], DAR 219.9: 135
To the editor of *The Times*, 23 July 1876, Harry Ransom Center, The Times (London, England) collection (*Correspondence* vol. 24)
To E. M. Dicey, [1877], DAR 202: 41 (*Correspondence* vol. 25)
From Charlotte Papé, 16 July 1875, DAR 174: 27 (*Correspondence* 23: 278)
From C. A. Kennard, 26 December 1881, DAR 210: 17 (*Correspondence* vol. 29)
To C. A. Kennard, 9 January 1882, DAR 185: 29–30 (*Correspondence* vol. 30)
From C. A. Kennard, 28 January 1882, DAR 185: 31 (*Correspondence* vol. 30)

Key to locations

APS: American Philosophical Society, Philadelphia, Pennsylvania, USA
Armacost Library: Armacost Library, University of Redlands, California, USA
Bodleian: Bodleian Library, University of Oxford, Oxford, UK
British Library: The British Library, Euston Road, London, UK
Bromley: Local Studies Library and Archives, Bromley Central Library, Bromley, Kent, UK
Brown University: John Hay Library, Brown University, Providence, Rhode Island, USA
Cleveland Health Sciences Library: Cleveland Health Sciences Library, Case Western Reserve University, Cleveland, Ohio, USA
CUL: Cambridge University Library, West Road, Cambridge, UK
DAR: Darwin Archive, Cambridge University Library, West Road, Cambridge, UK
Emma Darwin (1904) (publication): *Emma Darwin, wife of Charles Darwin. A century of family letters.* Edited by Henrietta Litchfield. 2 vols. Cambridge: privately printed by Cambridge University Press. 1904.
Field Museum: Field Museum of Natural History, Chicago, Illinois, USA
Harry Ransom Center: Harry Ransom Center, The University of Texas at Austin, Austin, Texas, USA
Houghton Library: Houghton Library, Harvard University, Cambridge, Massachusetts, USA
The Huntington Library: The Huntington Library, San Marino, California, USA
Imperial College: Imperial College of Science, Technology and Medicine Archives, South Kensington, London, UK
Lehigh: Lehigh University Libraries, Bethlehem, Pennsylvania, USA
Linnean Society of London Archives: Linnean Society of London, Piccadilly, London, UK
Liverpool Archives and Family History: Liverpool Archives and Family History, Central Library, William Brown Street, Liverpool, UK
Murray Archive: The John Murray Archive, National Library of Scotland, Edinburgh, UK
National Library of Australia: National Library of Australia, Canberra,

Australia

NHM: Natural History Museum, South Kensington, London, UK

North 1893 (publication): *Recollections of a happy life, being the autobiography of Marianne North*. 2d edition. 2 vols. Edited by Janet Catherine Symonds. London and New York: Macmillan and Co. 1893.

NYAM: New York Academy of Medicine, New York, New York, USA

R. M. Smythe (dealer): R. M. Smythe and Co., New York, New York, USA

Stargardt (dealer): J. A. Stargardt, Marburg, Germany

University of Chicago Library: University of Chicago Library, Chicago, Illinois, USA

University of Virginia Library: University of Virginia Library, Charlottesville, Virginia, USA

Vineland Historical and Antiquarian Society: Vineland Historical and Antiquarian Society, Vineland, New Jersey, USA

John Wilson (dealer): John Wilson Manuscripts, Cheltenham, UK

Yale, Beinecke: Yale University, Beinecke Rare Book and Manuscript Library, New Haven, Connecticut, USA

Biographical notes

This section contains brief biographies of the correspondents in this book, except for a few who could not be identified, and of many of the people mentioned in their letters and in the text. For more information, including biographical sources, see: www.darwinproject.ac.uk.

Abbot, Francis Ellingwood (1836–1903). American clergyman and philosopher. Helped found the Free Religious Association in 1867. Editor of the *Index* from 1870. President, National Liberal League, 1876–8. Helped found the National Liberal League of America after breaking with the NLL in 1878. Wrote on scientific theism.

Agassiz, Louis (1807–73). Swiss-born zoologist and geologist. Professor of natural history, Neuchâtel, 1832–46. Emigrated to the United States in 1846. Professor of zoology and geology, Harvard University, 1847–73. Established the Museum of Comparative Zoology at Harvard in 1859. A notable opponent of Darwin's theory of natural selection.

Alexander, Richard Chandler (1809–1902). Physician and botanist. Collected plants from around the world. Took the name Prior in 1859.

Allen, Fanny (1781–1875). Daughter of John Bartlett Allen and Elizabeth Allen. Emma Darwin's aunt.

Anderson, Elizabeth Garrett (1836–1917). Physician. Née Garrett; married James George Skelton Anderson in 1871. One of the first women to practise medicine in Britain. Obtained a medical licence from the Society of Apothecaries in 1865. MD, Paris, 1870. Set up in practice in London, and established the St Mary's Dispensary for Women and Children in 1866. Opened the New Hospital for Women (later the Elizabeth Garrett Anderson Hospital) in 1871. Taught at and sat on the council of the London School of Medicine for Women from 1874; dean, 1883–1902.

Aveling, Edward Bibbins (1851–98). Physician, socialist, and Karl Marx's son-in-law.

Barber, Mary Elizabeth (1818–99). British-born naturalist, artist, and writer in South Africa. Her family emigrated to South Africa in 1820. Married Frederick William Barber, a chemist, in 1845. Studied birds, moths, reptiles, and plants, and corresponded with leading scientists, providing them with specimens and drawings. Published a number of scientific papers.

Barnard, Anne (1833/4–99). Daughter of John Stevens Henslow and sister-in-law of J. D. Hooker. Married Robert Cary Barnard, an army officer, in 1859. Botanical artist; contributed plates to *Curtis's Botanical Magazine* and Daniel Oliver's *Lessons in elementary botany* (1864).

Bathoe, Maria Burnley (1817–85). Née Hume. Married Charles Gubbins (later Bathoe; d. 1866) of the Bengal Civil Service in 1843. Lived in India in the 1840s.

Becker, Lydia Ernestine (1827–90). Leading member of the women's suffrage movement, and botanist. Published *Botany for novices* (1864); awarded a Horticultural Society Gold Medal, 1865. Founder and president of the Manchester Ladies' Literary Society, 1867. Secretary to the Manchester National Society for Women's Suffrage from 1867. Member of the Manchester School Board, 1870. Editor of and regular contributor to the *Women's Suffrage Journal* from 1870. Secretary to the London central committee of the National Society for Women's Suffrage from 1881.

Bell, Charles (1774–1842). Anatomist and surgeon. Best known for his investigations of the nervous system and the expression of emotions in humans. Co-owner of and principal lecturer at the Great Windmill Street School of Anatomy, London, 1814–25. Surgeon at the Middlesex Hospital, 1812–36. Professor of surgery, Edinburgh University, 1836. Knighted, 1831. FRS 1826.

Bell, Marion (1787–1876). Daughter of Charles Shaw of Ayr. Married Charles Bell in 1811. Following the death of her husband in 1842, lived with her brother, Alexander Shaw; their house became a centre of literary and scientific society. Published her husband's letters in 1870.

Biddulph, Fanny Myddleton. *See* Mostyn Owen, Fanny.

Biddulph, Robert Myddleton. *See* Myddelton Biddulph, Robert.

Boner, Charles (1815–70). Journalist, translator, and poet. Tutored the two eldest sons of John Constable, 1831–7. Employed by the family of Prince Thurn und Taxis in Regensburg, Germany, 1840–60. Translated several German works, wrote verse, articles, and books; travelled as a naturalist and huntsman. Settled in Munich in 1860.

Bonham Carter, Elinor Mary. *See* Dicey, Elinor Mary.

Boole, Mary Everest (1832–1916). Writer and educator. Educated privately in France. In 1852, studied calculus with George Boole, whom she married in 1855. Librarian, Queen's College, London, 1865–73. Secretary to the doctor and philosopher James Hinton from 1873. Popular writer on mathematics and philosophy. Originator of Boole's Sewing Cards, an aid in teaching geometry. Many of her works focused on the psychology of learning.

Brodie, Jessie (1793–1873). Scottish children's nurse. Cared for the children of William Makepeace Thackeray, and for the Darwin children at 12 Upper Gower Street and Down House, 1842–51. Retired to Portsoy, Scotland, in 1851, after Annie Darwin's death.

Buckley, Arabella Burton (1840–1929). Popular scientific author, specialising in natural history. Secretary to Charles Lyell, 1864–75. Wrote in particular for young readers, and encouraged an active pursuit of natural history. A close friend of Alfred Russel Wallace and Darwin. A supporter of Darwinism; she emphasised the importance of mutualism and dependence as forces of evolution.

Buob, Louise (b. 1822/3). German-born governess and schoolmistress. Ran a private school at 2 Courtland Terrace, Kensington, London, circa 1862 – circa 1867.

Busk, Ellen (1816–90). Daughter of Jacob Hans Busk. Married her cousin, George Busk, in 1843.

Busk, George (1807–86). Russian-born naval surgeon and naturalist. Served on the hospital ship at Greenwich, 1832–55. Retired from medical practice in 1855. President of the Microscopical Society, 1848–9; of the Anthropological Institute, 1873–4. Zoological secretary of the Linnean Society of London, 1857–68. Hunterian Professor of comparative anatomy, Royal College of Surgeons of England, 1856–9. Specialised in palaeontology and in the study of Bryozoa. FRS 1850.

Butler, Josephine (1828–1906). Social reformer and women's activist. Née Grey. Married George Butler, an Anglican clergyman, in 1852. Cared for prostitutes and destitute women at her homes in Oxford and later Liverpool. Involved in campaigns for female education and suffrage. Published on women's position in society. Led the campaign against the Contagious Diseases Acts.

Butler, Mary (1810/11–66). Patient at Moor Park and Ilkley Wells hydropathic establishments. Visited the Darwins at Down. Sister of Richard Butler, vicar of Trim, Ireland.

Butler, Richard (1794–1862). Clergyman. BA, Oxford, 1817. Vicar of Trim, Ireland, 1819–62. A founder of the Irish Archaeological Society, 1840. Married Harriet Edgeworth, the sister of the novelist Maria Edgeworth, in 1826.

Canby, William Marriott (1831–1904). American botanist, businessman, and philanthropist. Lived in Wilmington, Delaware. Published several articles on insectivorous plants. Amassed a substantial herbarium.

Carlyle, Thomas (1795–1881). Essayist and historian.

Carpenter, Louisa (1812–87). Daughter of Joseph Powell, a merchant in Exeter. Married William Benjamin Carpenter in 1840.

Carpenter, William Benjamin (1813–85). Naturalist. Fullerian Professor of physiology at the Royal Institution of Great Britain, 1844–8; physiology lecturer, London Hospital, 1845–56; professor of forensic medicine, University College, London, 1849–59. FRS 1844.

Cecil, Sackville Arthur (1848–98). Company director. Fifth son of James Brownlow William Gascoyne-Cecil, second marquess of Salisbury (d. 1868), and Mary Catherine Gascoyne-Cecil, later Stanley. BA, Cambridge, 1869. Assistant general manager of the Great Eastern Railway, 1878–80. General Manager, Metropolitan District Railway, 1880–5. Chairman of the Exchange Telegraph Company, 1889–98.

Chambers, Robert (1802–71). Publisher, writer, and geologist. Partner, with his brother William Chambers, in the Edinburgh publishing company W. & R. Chambers. Joint editor of *Chambers's Edinburgh Journal* from 1832. Anonymous author of *Vestiges of the natural history of creation* (1844).

Cobbe, Frances Power (1822–1904). Irish-born writer and philanthropist. Wrote extensively on religious and ethical subjects. A leading campaigner for women's rights and against animal vivisection.

Congreve, Mary (1745–1823). Shrewsbury friend of the Darwin family. Sister of William Congreve, first baronet.

Congreve, William, 1st baronet (1743–1814). Comptroller of the Royal Laboratory, Woolwich, and superintendent of military machines. Created baronet, 1812.

Congreve, William, 2d baronet (1772–1828). Inventor of the Congreve rocket. MP for Plymouth, 1818–28. Succeeded his father as baronet in 1814. FRS 1811.

Craik, Georgiana Marion (1831–95). Novelist. Contributed short stories to Charles Dickens's *Household Words*.

Cresy, Edward (1824–70). Surveyor and civil engineer. Son of Edward Cresy (1792–1858), the architect and civil engineer who worked on alterations to Down House, Kent, for Darwin. Worked as an architectural draftsman in his father's office as a young man. Founder member of the Geologists' Association, 1858; president, 1864–5; vice-president, 1865–70.

Crookes, William (1832–1919). Chemist and science journalist. Superintendent of the meteorological department at the Radcliffe Observatory, Oxford, 1854; lecturer in chemistry, Chester Anglican teachers' training college, 1855. Editor of *Chemical News*, 1859–1906. Discovered the element thallium in 1861. Investigated mediums, including D. D. Home and Florence Cook, in the 1870s and 1880s. President of the Royal Society of London, 1913–15. Knighted, 1897. FRS 1863.

Cupples, Anne Jane (1839–98). Scottish author. Second daughter of Archibald Douglas. Married George Cupples in 1858. Wrote children's books. Lived in New Zealand from 1891.

Cupples, George (1822–91). Scottish writer and dog breeder. Served as an apprentice on an eighteen-month voyage to India and back on the *Patriot King, circa* 1838; had his indentures cancelled on his return. Studied arts and theology at Edinburgh University for eight years. Published a number of novels and other books, and wrote many articles and stories for journals. Bred Scottish deer-hounds.

Darwin, Amy Richenda (1850–76). Born in Wales. Daughter of Mary Anne and Lawrence Ruck. Married Francis Darwin, as his first wife, in 1874. Died shortly after the birth of their son, Bernard Richard Meirion Darwin.

Darwin, Anne Elizabeth (Annie) (1841–51). Charles and Emma Darwin's eldest daughter. Died at Malvern under the care of the physician and hydropathic doctor James Gully, possibly of tuberculosis.

Darwin, Bernard Richard Meirion (1876–1961). Son of Francis and Amy Darwin. Essayist and sports writer. Golf correspondent of *The*

Times, 1908–53. Played in the amateur golf championships and captained the British team in America in 1922. Captain of the Royal and Ancient Golf Club, 1934.

Darwin, Caroline Sarah. *See* Wedgwood, Caroline Sarah.

Darwin, Charles Robert (1809–82). Born in Shrewsbury, the grandson of Josiah Wedgwood I and Erasmus Darwin. Studied medicine in Edinburgh, 1825–7, but never practised. BA, Cambridge (Christ's College), 1831. Travelled on HMS *Beagle*, 1831–6. Married Emma Wedgwood, his cousin, in 1839. Moved to Down House, Down, Kent, in 1842. Published *On the origin of species by means of natural selection* in 1859.

Darwin, Charles Waring (1856–8). Youngest child of Charles and Emma Darwin. Died of scarlet fever.

Darwin, Elizabeth (1847–1926). Charles and Emma Darwin's third daughter to survive infancy. Went away to school for a brief period as a girl. Lived with her parents and after their deaths was head of her own household in Kensington and Cambridge.

Darwin, Emily Catherine (Catherine). *See* Langton, Emily Catherine (Catherine)

Darwin, Emma (1808–96). Youngest daughter of Josiah Wedgwood II. Married Charles Darwin, her cousin, in 1839. Bore ten children, eight of whom survived infancy. Took an active part in village affairs in Down, Kent, and after Charles's death divided her time between Down and Cambridge.

Darwin, Erasmus (1731–1802). Charles Darwin's grandfather. Physician, botanist, and poet. Advanced a theory of transmutation similar to that subsequently propounded by Jean Baptiste de Lamarck. FRS 1761.

Darwin, Erasmus Alvey (1804–81). Charles Darwin's brother. Attended Shrewsbury School, 1815–22. Matriculated at Christ's College, Cambridge, 1822; Edinburgh University, 1825–6. Qualified in medicine but never practised. Lived in London from 1829.

Darwin, Francis (1848–1925). Charles and Emma Darwin's third son. Botanist. BA, Cambridge, 1870. Qualified as a physician but did not practise. Charles's secretary from 1874. Collaborated with Charles on several botanical projects. Married Amy Ruck (d. 1876) in 1874 and Ellen Wordsworth Crofts in 1883. Lecturer in botany, Cambridge University, 1884; reader, 1888–1904. Published *Life and letters of Charles Darwin* and *More letters*. Knighted, 1913. FRS 1882.

Darwin, George Howard (1845–1912). Charles and Emma Darwin's second son. Mathematician. BA, Cambridge, 1868; fellow, 1868–78. Studied law in London, 1869–72; called to the bar, 1872, but did not practise. Plumian Professor of astronomy and experimental philosophy, Cambridge University, 1883–1912. Married Maud Du Puy of Philadelphia in 1884. Knighted, 1905. FRS 1879.

Darwin, Henrietta Emma. *See* Litchfield, Henrietta Emma.

Darwin, Horace (1851–1928). Charles and Emma Darwin's fifth son. Civil engineer. BA, Cambridge, 1874. Apprenticed to an engineering firm in Kent; returned to Cambridge in 1877 to design and make scientific instruments. Married Ida Farrer in 1880. Founder and director of the Cambridge Scientific Instrument Company. Mayor of Cambridge, 1896–7. Knighted, 1918. FRS 1903.

Darwin, Leonard (1850–1943). Charles and Emma Darwin's fourth son. Military engineer. Attended the Royal Military Academy, Woolwich. Commissioned in the Royal Engineers, 1871; major, 1889; retired, 1890. Served on several scientific expeditions, including those for the observation of the transit of Venus in 1874 and 1882. Instructor in chemistry and photography, School of Military Engineering, Chatham, 1877–82. Married Elizabeth Frances Frazer (d. 1898) in 1882, and Charlotte Mildred Massingberd in 1900.

Darwin, Marianne. *See* Parker, Marianne.

Darwin, Mary Eleanor (September–October 1842). Charles and Emma Darwin's third child.

Darwin, Robert Waring (1766–1848). Charles Darwin's father. Physician. Had a large practice in Shrewsbury and resided at The Mount. Son of Erasmus Darwin and his first wife, Mary Howard. Married Susannah, daughter of Josiah Wedgwood I, in 1796. FRS 1788.

Darwin, Sara Price Ashburner (1839–1902). Daughter of Sara Ashburner and Theodore Sedgwick. Sister of Susan Ridley Sedgwick Norton. Married William Erasmus Darwin in 1877.

Darwin, Susan Elizabeth (1803–66). Charles Darwin's sister. Lived at The Mount, Shrewsbury, the family home, until her death.

Darwin, William Erasmus (1839–1914). Charles and Emma Darwin's eldest son. Banker. BA, Cambridge, 1862. Partner in the Southampton and Hampshire Bank, Southampton, from 1861. Married Sara Price Ashburner Sedgwick in 1877.

Davies, Emily (1830–1921). Suffragist and promotor of higher education for women. Campaigned for secondary education for girls, women's suffrage, and higher education for women. Helped found Girton College, Cambridge.

Derby, Lady. *See* Stanley, Mary Catherine.

Dicey, Albert Venn (1835–1922). Jurist. BA, Oxford, 1858. Journalist and scholar in London, 1861–82. Called to the bar, 1863. Married Elinor Mary Bonham-Carter in 1872. Involved in the foundation of Newnham College, Cambridge, a women's college.

Dicey, Elinor Mary (1837–1923). Daughter of Joanna Maria Bonham-Carter, a family friend of the Darwins. The family lived at Keston, Kent, from 1853. Married Albert Venn Dicey in 1872. Involved in the foundation of Newnham College, Cambridge, a women's college.

Dillwyn Llewelyn, Thereza. *See* Story-Maskelyne, Thereza.

Dixie, Florence Caroline (1855–1905). Scottish traveller and writer. Daughter of Archibald William Douglas, eighth marquess of Queensbury, and his wife, Caroline Margaret. Married Sir Alexander Beaumont Churchill Dixie in 1875. Travelled to South America in 1878 and published *Across Patagonia* (1880). Appointed war correspondent in South Africa by the *Morning Post*; spent six months in southern Africa, publishing articles in the *Morning Post* and two further books. Advocated complete sex equality and, later in life, denounced the cruelty of blood sports and advocated secularism.

Doedes, Nicolaas Dirk (1850–1906). Dutch topographical writer. Son of a theologian; studied theology at the University of Utrecht before changing to history.

Dowie, Annie (1835–1903). Daughter of Robert and Anne Kirkwood Chambers. Married James Muir Dowie.

Drysdale, Elizabeth (1781/2–1882). Daughter of John Pew of Hilltown, Kirkudbrightshire. Married William Copland of Colliston, Dumfries; widowed, 1808. Married Sir William Drysdale (1781–1847), for many years treasurer of the city of Edinburgh. Mother of John James Drysdale, a leading homeopathic doctor. Mother-in-law of the hydropathic specialist Edward Wickstead Lane.

Eliot, George (1819–80). Novelist. Born Mary Ann Evans. Translated works of liberal theology from German into English. Worked as a journalist in London with John Chapman. Lived with G. H. Lewes from 1854. Published her first novel, *Scenes of clerical life*, under the pseudonym George Eliot, in 1858.

Engleheart, Stephen Paul (1831/2–85). Surgeon. Surgeon in Down, Kent, 1861–70. Medical officer, Second District, Bromley Union, 1863–70; divisional surgeon of police, 1863–70. Resident in Shelton, Norfolk, 1870–81; in Old Calabar, Nigeria, 1882–5.

Farrer, Emma Cecilia (Ida) (1854–1946). Only daughter of Thomas Henry Farrer and Frances Farrer (née Erskine); distantly related to Charles and Emma Darwin. When her father remarried, became step-daughter of Katherine Euphemia (Effie) Wedgwood. Married Horace Darwin in 1880, and with him built, in 1884, and resided at The Orchard, Cambridge. Active in Cambridgeshire charities related to mental health.

Farrer, Katherine Euphemia (Effie) (1839–1931). Daughter of Hensleigh and Frances Emma Elizabeth Wedgwood. Married Thomas Henry Farrer as his second wife in 1873.

Fawcett, Henry (1833–84). Economist and politician. BA, Cambridge, 1856. Blinded in a shooting accident in 1858. Appointed professor of political economy, Cambridge University, 1863. Sought to open Cambridge University to all classes and religious groups. MP for Brighton, 1865–74; Hackney, 1874–84. Shared John Stuart Mill's enthusiasm for feminism. Married Millicent Garrett Anderson in 1867. FRS 1882.

Fawcett, Millicent Garrett (1847–1929). Leader of the constitutional women's suffrage movement and author. Sister of Elizabeth Garrett Anderson. Married Henry Fawcett in 1867. Wrote on women's education, women's suffrage, and economics, as well as two novels. Co-founder of Newnham College in 1875, and served on its council. Committee member of the London National Society for Women's Suffrage from 1867. Spoke and lectured on women's issues and other political subjects in the 1870s. Emerged as the women's suffrage movement's leader after the death of Lydia Becker in 1890. Honorary LLD, University of St Andrews, 1899. President of the National Union of Women's Suffrage Societies, 1907–19. Dame Grand Cross of the British Empire, 1925.

Fawcett, Philippa Garrett (1868–1948). Mathematician and civil servant. Daughter of Henry and Millicent Garrett Fawcett. Studied at Newnham College, Cambridge, and received the highest marks in the Mathematical Tripos in 1890. Lecturer, Newnham College, 1892–1902. Principal assistant in the executive officer (education) department of London County Council, 1905–20; assistant education officer (higher education), 1920–34.

Ffinden, George Sketchley (1836/7–1911). Clergyman. Ordained priest, 1861. Vicar of Down, 1871–1911.

Forster, Laura May (1839–1924). A life-long friend of Henrietta Emma Darwin, whom she met while staying in Wales in 1865. An aunt of E. M. Forster. Made observations on harvester ants in Algiers for John Traherne Moggridge in 1872. Forster loaned her house to the Darwins in 1879 so that Charles Darwin could have a complete rest.

Galton, Francis (1822–1911). Traveller, statistician, and scientific writer. Charles Darwin's cousin. Explored in south-western Africa, 1850–2. Carried out various researches on heredity. Founder of the eugenics movement. FRS 1860.

Glen, Catherine (b. *c.* 1820). Scottish teacher. Daughter of William Glen (1789–1826), poet, and Catherine (Kate) Glen of Aberfoyle, Stirling. With her mother, cared for and trained children from the Glasgow poorhouse at Aberfoyle.

Gourlay, Jane. Schoolteacher in Edinburgh. Friend of Frances Julia Wedgwood. Tutor to the Paterson children at Linlathen, Forfarshire, the estate of Thomas Erskine (1788–1870), theologian.

Gray, Asa (1810–88). American botanist. Fisher Professor of natural history, Harvard University, 1842–73. Wrote numerous botanical textbooks and works on North American flora. President of the American Academy of Arts and Sciences, 1863–73; of the American Association for the Advancement of Science, 1872. Regent of the Smithsonian Institution, 1874–88. Foreign member, Royal Society of London, 1873.

Gray, Jane Loring (1821–1909). Daughter of Charles Greely Loring, Boston lawyer and politician, and Anna Pierce Brace. Married Asa Gray in 1848. Edited the *Letters of Asa Gray* (1893).

Gully, James Manby (1808–83). Physician. Practised medicine in London, 1830–42. Set up a hydropathic establishment in Great Malvern, Worcestershire, in 1842; a successful practitioner of hydropathy until his retirement in 1872.

Haeckel, Ernst (1834–1919). German zoologist. Professor extraordinarius of zoology, Jena, 1862–5; professor of zoology and director of the Zoological Institute, 1865–1909. Specialist in marine invertebrates. Leading populariser of evolutionary theory. His *Generelle Morphologie der Organismen* (1866) linked morphology to the study of the phylogenetic evolution of organisms.

Haliburton, Sarah Harriet (1810/11–86). Eldest daughter of William Mostyn Owen Sr of Woodhouse. Married Edward Hosier Williams (d. 1844) in 1831 and Thomas Chandler Haliburton (d. 1865) in 1856. A close friend and neighbour of Darwin's before the *Beagle* voyage.

Harrison, Lucy Caroline. *See* Wedgwood, Lucy Caroline.

Hawkshaw, Ann (1812–85). Poet. Daughter of the Rev. James Jackson of Green Hammerton, Yorkshire. Married John Hawkshaw in 1835. Published several volumes of poetry including *Sonnets on Anglo-Saxon history* and *Cecil's own book*, which was for her grandson, Cecil Wedgwood.

Hawkshaw, Cicely Mary (1837–1917). Daughter of Francis and Frances Wedgwood. Emma Darwin's niece. Married Clarke Hawkshaw in 1865.

Hawkshaw, John (1811–91). Civil engineer. In charge of the Bolivar Mining Association's mines in Venezuela, 1832–4. Engineer to the Manchester and Leeds Railway, 1845–88. Practised as a consulting engineer in London from 1850. One of the foremost civil engineers of the nineteenth century. Knighted, 1873. FRS 1855.

Henslow, Harriet (1797–1857). Daughter of George Leonard Jenyns and sister of Leonard Jenyns. Married John Stevens Henslow in 1823.

Henslow, John Stevens (1796–1861). Clergyman, botanist, and mineralogist. Darwin's teacher and friend. Professor of mineralogy, Cambridge University, 1822–7; professor of botany, 1825–61. Extended and remodelled the Cambridge botanic garden. Curate of Little St Mary's Church, Cambridge, 1824–32; vicar of Cholsey-cum-Moulsford, Berkshire, 1832–7; rector of Hitcham, Suffolk, 1837–61.

Herrick, Sophie Bledsoe (1837–1919). American editor and writer. Head of a Baltimore girls' school, 1868–72. Writer and associate editor, *Southern Review*, 1872–77; editor, 1877. Studied biology at Johns Hopkins in 1876. An editor of *Scribner's Monthly* from 1879, and its successor, *Century*, until 1906. Wrote science books for children, mostly on botany and geology.

Hill, John (1802–91). Clergyman. BA, Oxford, 1824. Rector of Great Bolas, Shropshire, 1831–77.

Hooker, Frances Harriet (1825–74). Daughter of John Stevens Henslow. Married Joseph Dalton Hooker in 1851. Assisted her husband significantly in his published work. Translated *A general system of botany, descriptive and analytical*, by Emmanuel Le Maout and Joseph Decaisne, from the French (1873).

Hooker, Joseph Dalton (1817–1911). Botanist. Worked chiefly on taxonomy and plant geography. Friend and confidant of CD. Accompanied James Clark Ross on his Antarctic expedition, 1839–43, and published the botanical results of the voyage. Travelled in the Himalayas, 1847–9. Assistant director, Royal Botanic Gardens, Kew, 1855–65; director, 1865–85. Knighted, 1877. FRS 1847.

Horner, Leonard (1785–1864). Scottish geologist and educationalist. Founded the Edinburgh School of Arts in 1821. Warden of University College, London, 1827–31. Inspector of factories, 1833–59. A promoter of science-based education at all social levels. President of the Geological

Society of London, 1845–7 and 1860–2. Father-in-law of Charles Lyell. FRS 1813.

Humboldt, Alexander von (1769–1859). Prussian naturalist, geographer, and traveller. Explored northern South America, Cuba, Mexico, and the United States, 1799–1804. Travelled in Siberia in 1829. Foreign member, Royal Society of London, 1815.

Huxley, Henrietta Anne (1825–1915). Born Henrietta Anne Heathorn. Emigrated to Australia in 1843. Met Thomas Henry Huxley in Sydney, Australia, in 1847, and married him in 1855. Translated German for him and drew diagrams for his lectures. The couple had eight children.

Huxley, Thomas Henry (1825–95). Zoologist. Assistant-surgeon on HMS *Rattlesnake*, 1846–50, during which time he investigated Hydrozoa and other marine invertebrates. Lecturer in natural history, Royal School of Mines, 1854; professor, 1857. Appointed naturalist to the Geological Survey of Great Britain, 1854. Hunterian Professor, Royal College of Surgeons of England, 1862–9. Fullerian Professor of physiology, Royal Institution of Great Britain, 1855–8, 1866–9. FRS 1851.

Innes, John Brodie (1815–94). Clergyman. Perpetual curate of Down, 1846–68; vicar, 1868–9. Left Down in 1862 after inheriting an entailed estate at Milton Brodie, near Forres, Scotland; changed his name to Brodie Innes in 1861 as required by the entail.

Isitt, Virginia Lavinia (1837–88). Teacher. Governess to the children of Emily Jesse, Alfred Tennyson's sister. Studied for the French State Certificate in teaching at the Convent College in Arras, France, 1862–4. First headmistress of the Port Elizabeth Collegiate School for girls, in South Africa, 1874–86.

Johnson, Henry (1802/3–81). Physician. A contemporary of Darwin's at Shrewsbury School and Edinburgh University. Senior physician, Shropshire Infirmary. Member of Royal College of Physicians of London, 1859. Founder member and honorary secretary of the Shropshire and North Wales Natural History and Antiquarian Society, 1835–77.

Kempson, Louisa Frances (1834–1903). Daughter of Henry Allen and Jessie Wedgwood. Emma Darwin's niece. Married William John Kempson in 1864.

Kennard, Caroline Augusta (1827–1907). American feminist. Née Smith. Married Martin Parry Kennard in 1847; he was a businessman in Boston, Mass., and an anti-slavery activist. They moved to Brookline, Mass., in 1854. Interested in the botany of ferns and mosses. Published a biography of Dorothea Dix (1888). A science scholarship was established in her name at Radcliffe College by her sister, Martha T. Fiske Collord.

Kettle, Rosa Mackenzie (d. 1895). Novelist and poet. Born Mary Rosa Stuart Kettle, but adopted the name Mackenzie, her mother's maiden name.

Lane, Edward Wickstead (1823–89). Physician. Proprietor of a hydropathic establishment at Moor Park, near Farnham, Surrey, 1859 (or before), and at Sudbrook Park, near Ham, Surrey, 1860–79. Practised in

Harley Street, London, 1879–89. Member of the Faculty of Advocates, the Botanical Society, and the Speculative Society, Edinburgh. Author of works on hydropathy.

Lane Fox, Alice Margaret (1828–1910). Eldest daughter of Edward John Stanley, second baron of Alderley. Married Augustus Henry Lane Fox (later Pitt-Rivers) in 1853.

Langton, Charles (1801–86). Rector of Onibury, Shropshire, 1832–41. Left the Church of England in 1841. Resided at Maer, Staffordshire, 1841–7, and at Hartfield Grove, Hartfield, Sussex, 1847–63. Married Emma Darwin's sister, Charlotte Wedgwood, in 1832. After her death, married Charles Darwin's sister, Emily Catherine Darwin, in 1863.

Langton, Charlotte (1797–1862). Emma Darwin's sister. Married Charles Langton in 1832. Resided at Maer, Staffordshire, 1840–6, and at Hartfield Grove, Hartfield, Sussex, 1847–62.

Langton, Emily Catherine (Catherine) (1810–66). Charles Darwin's sister. Resided at The Mount, Shrewsbury, until she married Charles Langton in 1863.

Lewes, George Henry (1817–78). Writer. Author of a biography of Goethe (1855). Contributed articles on literary and philosophical subjects to numerous journals. Editor, *Fortnightly Review*, 1865–6. Published on physiology and on the nervous system in the 1860s and 1870s. Lived with Marian Evans (George Eliot) from 1854.

Litchfield, Henrietta Emma (1843–1927). Charles and Emma Darwin's second daughter to survive infancy. Married Richard Buckley Litchfield in 1871. Assisted her father with his publications. Edited *Emma Darwin* (1904) and (1915).

Litchfield, Richard Buckley (1832–1903). Barrister. BA, Cambridge, 1853. Admitted to the Inner Temple, 1854; called to the bar, 1863. First-class clerk in the office of the Ecclesiastical Commissioners. Married Henrietta Emma Darwin in 1871. A founder and treasurer of the Working Men's College; taught mathematics there, 1854–70, and music from 1860.

Lloyd, Mary Charlotte (1819–96). Welsh landowner and artist. Studied in Italy with the sculptor John Gibson. Met Frances Power Cobbe there and became her lifelong companion. Neighbour of the Darwin family during their stay at Caerdeon, Barmouth, Wales, in summer 1869.

Lubbock, Ellen Frances (1834/5–79). Daughter of the Rev. Peter Hordern of Chorlton-cum-Hardy, Lancashire. Married John Lubbock in 1856. Reviewed Darwin's *Insectivorous plants* in *Academy*. Wrote about her visit with her husband to the Danish shell-mounds in Francis Galton's *Vacation tourists*. Acted as a scientific hostess and secretary for her husband.

Lubbock, John, 4th baronet and 1st Baron Avebury (1834–1913). Banker, politician, and naturalist. Son of John William Lubbock and a neighbour of Darwin's in Down. Studied entomology and anthropology. Worked at the family bank from 1849; head of the bank from 1865. Liberal MP for Maidstone, Kent, 1870–80; for London University,

1880–1900. Succeeded to the baronetcy in 1865. Created Baron Avebury, 1900. FRS 1858.

Ludwig, Camilla (1837/8–1912). Born in Hamburg. Governess to the Darwin family, 1860–3. Translated German works for Darwin. Married Reginald Saint Pattrick, vicar of Sellinge, Kent, in 1874.

Ludwig, Louisa (1839/40–1915). Born in Hamburg. Governess to the Darwin family for periods in the 1860s. Sister of Camilla Ludwig.

Lyell, Charles, 1st baronet (1797–1875). Scottish geologist. Uniformitarian geologist whose *Principles of geology* (1830–3), *Elements of geology* (1838), and *Antiquity of man* (1863) appeared in many editions. Professor of geology, King's College, London, 1831. Travelled widely and published accounts of his trips to the United States. Darwin's scientific mentor and friend. Knighted, 1848; created baronet, 1864. FRS 1826.

Lyell, Henry (1804–75). Army officer in India. Married Katharine Murray Horner in 1848. Brother of Charles Lyell.

Lyell, Katharine Murray (1817–1915). Botanist and literary editor. Daughter of Leonard Horner. Married Henry Lyell, brother of Charles Lyell, in 1848. Her Indian plant collections are at the British Museum, and her fern collection was given to Kew. Edited the lives and letters of Charles Lyell and Charles James Fox Bunbury,

Lyell, Mary Elizabeth (1808–73). Eldest child of Leonard Horner. Married Charles Lyell in 1832. Linguist and conchologist.

Mackintosh, Frances Emma Elizabeth (Fanny). *See* Wedgwood, Frances Emma Elizabeth (Fanny).

Mackintosh, James (1765–1832). Philosopher and historian. Professor of law and general politics at the East India Company College, Haileybury, 1818–24. Married Catherine Allen, his second wife, sister-in-law of Josiah Wedgwood, in 1798. Knighted, 1803.

Marshall, Theodosia (1841–92). Daughter of Henry Cowper Marshall and sister of William Cecil Marshall. Member of the Leeds Ladies' Educational Association; gave lectures on political economy. Sent Darwin observations on insectivorous plants.

Marshall, William Cecil (1849–1921). Architect. BA, Cambridge, 1872. Designed many tennis courts, and the billiard room with dressing room and bedroom above for Down House.

Martineau, Harriet (1802–76). Author, reformer, and traveller. Friend of Erasmus Alvey Darwin and Hensleigh and Frances Emma Elizabeth Wedgwood. Her *Illustrations of political economy* (1832), twenty-three tales illustrating economic principles, was her first popular success.

Meteyard, Eliza (1816–79). Author. Contributed fiction and social articles to numerous periodicals under the pen-name Silverpen. Her novels include *Struggles for fame* (1845). An active member of the Whittington Club from 1846. Published a number of works about the Wedgwoods including a two-volume life of Josiah Wedgwood I (1865–6), and *The Wedgwood handbook* (1875).

Michelet, Adèle-Athénaïs Mialaret (Athénaïs) (1826–99). French author and natural historian. Taught in Vienna. Married Jules Michelet as his second wife in 1849. Collaborated in his work on *L'Oiseau*, *L'Insecte*, and *La Mer*, and was his literary executor. Published natural history and other works in her own right.

Michelet, Jules (1798–1874). French historian and writer on natural history. Director of the historical section of the national archives from 1831; professor of history and moral philosophy, Collège de France, from 1838. Married Athénaïs Mialaret as his second wife in 1849. Author of *Histoire de France* and numerous other works.

Mill, John Stuart (1806–73). Philosopher, political economist, and advocate of women's rights. MP for Westminster, 1865–8. In 1867, forced a debate on an amendment to Disraeli's suffrage bill, proposing the removal of terms limiting the franchise to males. In 1868, presented a petition demanding amendment of the law regarding married women's property. Published *The subjection of women* (1869).

Morley, John, Viscount Morley of Blackburn (1838–1923). Politician and writer. Lost his faith while at Lincoln College, Oxford; earned a living in London from 1860 as a teacher and journalist. Editor of the *Fortnightly Review*, 1867–82; of the *Pall Mall Gazette*, 1880–3; of *Macmillan's Magazine*, 1883–5. Liberal MP for Newcastle, 1883–95; for Montrose Burghs, 1896–1908. Created Viscount Morley of Blackburn, 1908.

Morris, Margaretta Hare (1797–1867). American entomologist. Of Philadelphia, Pa. Published studies of the Hessian fly and the seventeen-year locust.

Mostyn Owen, Fanny (1806/7–87). Second daughter of William Mostyn Owen. Married Robert Myddelton Biddulph in 1832. A close friend and neighbour of Darwin's before the *Beagle* voyage.

Mostyn Owen, Harriet Elizabeth. Wife of William Mostyn Owen.

Mostyn Owen, Sarah Harriet. *See* Haliburton, Sarah Harriet.

Mostyn Owen, William (1769/70–1849). Army officer. Lieutenant, Royal Dragoons. Squire of Woodhouse, Shropshire. Father of Fanny and Sarah Harriet Mostyn Owen, amongst other sons and daughters.

Murray, John (1808–92). Darwin's publisher from 1845.

Myddelton Biddulph, Fanny. *See* Mostyn Owen, Fanny.

Myddelton Biddulph, Robert (1805–72). Politician. Whig MP for Denbighshire, 1832–5 and 1852–68. Colonel of the Denbighshire Militia, 1840–72. Lord lieutenant of Denbighshire, 1841–72. Aide-de-camp to Queen Victoria, 1869–72. Married Fanny Mostyn Owen in 1832.

Nash, Louisa A'hmuty (1838–1922). Author. Daughter of Henry and Mary Desborough. Married Wallis Nash in 1868; they lived in Down during part of the 1870s, and later emigrated to the US. Painted a portrait of Darwin in indian ink. Author of *Recollections of Abraham Lincoln* (1897).

Nash, Wallis (1837–1926). Lawyer and agriculturalist. Studied at New College, University of London. Lived at The Rookery, north of Down, Kent, 1873–7. Emigrated to Oregon in 1879. Practised law and farming. Involved in founding the Oregon Pacific Railroad and Oregon Agricultural College. Editorial writer for the *Oregon Journal*. Wrote about his travels in Oregon.

Nevill, Dorothy Fanny (1826–1913). Society hostess and horticulturist. Daughter of Horatio Walpole, third earl of Orford; married her cousin Reginald Henry Nevill in 1847. Developed a notable garden at Dangstein, near Petersfield, Hampshire, where she cultivated orchids, pitcher-plants, and other tropical plants; employed thirty-four gardeners. Sent Darwin specimens of rare orchids and insectivorous plants.

Nevill, Reginald Henry (1807–78). Son of George Henry Nevill and Caroline Walpole. Married his cousin Dorothy Fanny Nevill in 1847.

North, Marianne (1830–90). Painter and traveller. In the 1870s, visited Canada, the US, Jamaica, Brazil, California, Japan, Borneo, Java, Ceylon (Sri Lanka), and India. Presented her collection of paintings to the Royal Botanic Gardens, Kew, for display in a gallery designed, furnished, and financed by herself. Visited Australia and New Zealand in 1880, South Africa and the Seychelles in 1882–3, and Chile in 1884–5.

Norton, Charles Eliot (1827–1908). American editor, literary critic, and art historian. Graduated from Harvard College in 1846. Apprenticed himself in the East India trade, travelling widely in India and Europe. Contributed to the *Atlantic Monthly*; co-edited the *North American Review*, 1863–8; and co-founded and wrote for the *Nation*. Travelled and lived in England and continental Europe, 1868–73. Taught history of art and literature at Harvard, 1874–98.

Norton, Susan Ridley Sedgwick (1838–72). Daughter of Sara Ashburner and the American legal theorist Theodore Sedgwick. Grew up in New York and Massachusetts. Married Charles Eliot Norton in 1862. Died in Dresden, Germany, after giving birth to her sixth child.

Olmsted, Frederick Law (1822–1903). American landscape architect. Appointed superintendent of Central Park, New York City, 1857; architect-in-chief, 1858. Superintendent, Frémont Mariposa mining estates, California, 1863–5. Reappointed landscape architect, Central Park, 1865, until his removal in 1878. Thereafter worked independently.

Owen, Fanny Mostyn. *See* Mostyn Owen, Fanny.

Owen, Harriet Elizabeth Mostyn. *See* Mostyn Owen, Harriet Elizabeth.

Owen, Richard (1804–92). Comparative anatomist. Hunterian Professor of comparative anatomy and physiology, 1836–56. Superintendent of the natural history departments, British Museum, 1856–84; prime mover in establishing the Natural History Museum, South Kensington, 1881. Described the *Beagle* fossil mammal specimens, but later became a bitter personal enemy of Darwin's. Knighted, 1884. FRS 1834.

Owen, Sarah Harriet Mostyn. *See* Haliburton, Sarah Harriet.

Owen, William Mostyn. *See* Mostyn Owen, William.

Papé, Charlotte (*fl.* 1870s). Fluent in English and German; interested in women's rights. In Manchester, 1875. Probably the author of an article about mothers' legal rights over their children in a German women's journal, 1876. In Leipzig, planning to publish an appreciation of John Stuart Mill in a German women's journal, 1879.

Parker, Marianne (1798–1858). Darwin's eldest sister. Married Henry Parker (1788–1856) in 1824.

Parslow, Joseph (1811/12–98). Darwin's manservant at 12 Upper Gower Street, London, *circa* 1840–2, and butler at Down House until 1875.

Perfil'eva, Praskov'ja Fëdorovna (Pauline Perfilieff) (1831–87). Russian. Née Tolstaya. Daughter of Fëdor Ivanovich Tolstoy, the second cousin (once removed) of Count Lev Nikolaevich Tolstoy (Leo Tolstoy). Married Vasilij Stepanovich Perfil'ev. Informally known as Polen'ka; signed herself Pauline Perfilieff.

Pfeiffer, Emily (1827–90). Welsh-born poet. Née Davis. Married Jurgen Edward Pfeiffer, a German tea merchant resident in London, in 1850. Published several works of poetry, beginning in 1857, including a volume on her travels in eastern Europe, Asia, and the United States in 1884. Interested in the position of women in society.

Powell, Henry (1839/40–92). Clergyman. BA, Cambridge, 1861. Ordained priest, 1864. Vicar of Down, 1869–71.

Pugh, Mary Ann (1824/5–95). Governess to the Darwin children, 1857–9. Previously governess to the children of John and Ann Hawkshaw. Later became insane and lived at the Priory, Roehampton, a private asylum.

Rich, Mary (1789–1876). Daughter of Sir James Mackintosh. Half-sister of Frances Emma Elizabeth Wedgwood. Married Claudius James Rich in 1808.

Riley, Charles Valentine (1843–95). Entomologist. Emigrated to the United States *circa* 1859. State entomologist of Missouri, 1868–76. Became a US citizen in 1869. Chief of the Department of the Interior's US Entomological Commission, 1877–82. Entomologist with the Division of Entomology of the Department of Agriculture, 1878, 1881–94.

Robinson, John Warburton (b. 1837/8). Clergyman. Curate of Down, 30 August 1868 to 4 February 1869.

Ruck, Amy Richenda. *See* Darwin, Amy Richenda.

Ruck, Lawrence (1819/20–96). Landowner. Married Mary Anne Matthews in 1841. Father of Amy Richenda Ruck, who married Francis Darwin in 1874. Was committed to an insane asylum in the 1850s, but later released.

Ruck, Mary Anne (1821/2–1905). Née Matthews; married Lawrence Ruck in 1841. Lived at Pantlludw, near Machynlleth, Wales. Mother of Amy Ruck, who married Francis Darwin in 1874.

Salwey, Charlotte Margaretta (d. 1858). Married Richard Betton of Overton House, Shropshire, in October 1831.

Sargent, Charles Sprague (1841–1927). American botanist. Graduated from Harvard in 1862. Served in the Union Army during the American Civil War. Travelled in Europe, 1865-8. Director of the Harvard Botanic Garden, 1872–3; first director of the Arnold Arboretum, Harvard University, 1873. Arnold Professor of arboriculture at Harvard, 1879–1927.

Sedgwick, Sara Price Ashburner. *See* Darwin, Sara Price Ashburner.

Shaw, Alexander (1804–90). Scottish surgeon. MA, University of Glasgow, 1822. Entered Middlesex Hospital, London, as a pupil, 1822; assistant surgeon, 1836, surgeon, 1842. Published on and edited works by Sir Charles Bell, his brother-in-law.

Shuttleworth, Caroline (1835/6–1918). Daughter of Emma Martha Shuttleworth and Philip Nicholas Shuttleworth, warden of New College, Oxford, and bishop of Chichester. Sister of the translator and poet Frances Bevan.

Shuválov (or Schouvaloff), Peter Andreivich, Count (1827–89). Russian diplomat. Ambassador to London during the 1877–8 Russo-Turkish war.

Sidgwick, Henry (1838–1900). Philosopher. BA, Cambridge, 1859; fellow and lecturer in classics, 1859–69; lecturer, 1869–83; Knightbridge Professor of moral philosophy, 1883–1900. Author of *The methods of ethics* (1874). First president of the Society for Psychical Research, 1882–5. Promoter of the higher education of women.

Snow, Maria Isabella (1851/2–79). Daughter of the Rev. George D'Oyly Snow and Maria Jane Snow. Of Down Wood House, Langton Long Blandford, Dorset.

Somerville, Mary (1780–1872). Writer on science and mathematics. Author of a condensed English version of Laplace's *Mécanique céleste* (1831), and of *On the connection of the physical sciences* (1834). Advocate of higher education for women and women's suffrage.

Stanley, Edward Henry, 15th earl of Derby (1826–93). Politician and diarist. BA, Cambridge, 1848. MP for King's Lynn, 1848–69. Visited the West Indies twice, 1848–50. First secretary of state for India from 1858. Foreign secretary from 1874. Colonial secretary from 1882. Succeeded to the earldom in October 1869.

Stanley, Mary Catherine, countess of Derby (1824–1900). Political hostess. Daughter of George Sackville-West, fifth Earl De La Warr. Married James Brownlow William Gascoyne-Cecil in 1847. After his death, married Edward Henry Stanley, fifteenth earl of Derby, in 1870. Deeply involved in Conservative politics.

Stansfeld, Caroline (1816–85). Radical. Her father, the lawyer William Henry Ashurst, was progressive in his views on women's rights. Married James Stansfeld, a radical lawyer, in 1844. Worked with the Associate Institution (campaigning for the reform of laws on prostitution), the Whittington Club, and the London Society for Women's Suffrage, and campaigned against the Contagious Diseases Act. Also worked for the unification of Italy.

Stansfeld, James (1820–98). Politician and social reformer. A supporter of the Northern Reform Union. Elected radical MP for Halifax, 1859. Financial secretary to the Treasury, 1869; president of the Local Government Board, 1871; he appointed the first female poor-law inspector in 1872. Knighted, 1895.

Stephen, Caroline Emelia (1834–1909). Author. Younger sister of Leslie Stephen. Became a Quaker. Cared for her mother in her last illness and for Leslie after the death of his first wife. Published *The service of the poor* (1871), a study of religious sisterhoods, *Quaker strongholds* (1891), *Light arising* (1908), and *The vision of faith* (1911).

Stephens, Thomas Sellwood (b. 1825). Clergyman. Curate at Down, Kent, 1859–67.

Stokes, George Gabriel, 1st baronet (1819–1903). Physicist. Lucasian Professor of mathematics, Cambridge University, 1849–1903. Secretary of the Royal Society of London, 1854–85; president, 1885–90. Conservative MP for Cambridge University, 1887–91. Created baronet, 1889. FRS 1851.

Story-Maskelyne, Nevil (1823–1911). Mineralogist. Lectured on mineralogy, Oxford University, 1850–7; professor of mineralogy, 1856–95. Keeper of the mineral department, British Museum, 1857–80. MP for Cricklade, 1880–5; North Wiltshire, 1885–92. FRS 1870.

Story-Maskelyne, Thereza (1834–1926). Welsh botanist, astronomer, and experimental photographer. A granddaughter of the naturalist Lewis Weston Dillwyn, and daughter of the photographer John Dillwyn Llewelyn. Supplied climate data to the British Association for the Advancement of Science. Married Nevil Story-Maskelyne in 1858. Mother of three daughters, including the educator and gardener Thereza, Lady Rucker.

Tait, Lawson (1845–99). Scottish gynaecological surgeon. Started a practice in Birmingham in 1870. Junior surgeon, Birmingham and Midland Hospital for Women, 1871. Instigated a nurse's training programme, and supported education and professional positions for women as nurses and doctors. Internationally recognised pioneer in abdominal surgery, especially ovariotomy. Professor of gynaecology, Queen's College, Birmingham, 1887.

Tait, Sybil Anne (1844–1909). Daughter of William Stewart. Married Lawson Tait in 1871. Member of the Birmingham Natural History Society and the Birmingham Philosophical Society.

Talbot, Emily Fairbanks (1834–1900). American philanthropist and promoter of higher education for women. A teacher in Baltimore from 1854. Married Israel Tisdale Talbot, a homoeopathic doctor, in 1856; promoted and secured funds for his work in Boston. In 1877, helped to establish the Boston Latin School for Girls, which offered a college preparatory course for female students. Helped form the Association of Collegiate Alumnae, to improve, promote, and set standards for women's higher education, in 1881. Secretary of the education department of the American Social Science Association.

Tanner, Mary Willes (1836–1916). Daughter of John Roberts, professor of music, and Charlotte Elizabeth Roberts. Married Thomas Hawkes Tanner, physician, in 1859.

Taylor, Helen (1831–1907). Promoter of women's rights. Involved, along with her stepfather, John Stuart Mill, in the women's suffrage cause, and acted as his amanuensis. In 1876, elected to Southwark school board; was twice re-elected but resigned in 1884 on grounds of ill health. Supported causes such as free education, Irish home rule, land reform, and the banning of the Contagious Diseases Acts and fox-hunting.

Tennyson, Alfred, 1st Baron Tennyson (1809–92). Poet laureate, 1850. Created Baron Tennyson, 1883.

Thackeray, Anne Isabella (1837–1919). Writer. Eldest surviving daughter of William Makepeace Thackeray. Acted as her father's amanuensis, and published novels and reminiscences. Married her cousin Richmond Thackeray Willoughby Ritchie in 1877.

Thom, J. P. (b. 1838/9). Journalist. Patient at Edward Wickstead Lane's hydropathic establishment in the 1850s. Subeditor of the colonial newspaper *Home News*. Emigrated to Queensland, Australia, in 1863.

Thorley, Catherine (1828/9–1911). Governess at Down House, 1850–6. Present at Annie Darwin's death in Malvern in 1851.

Thorley, Elizabeth Mary (1805/6–59). Mother of Catherine and Emily Thorley, governesses to the Darwin family. Friend of the Tollet family. Widowed and residing at 36 Bernard Street, Russell Square, London, in 1851.

Thorley, Emily Maria (1832/3–1917). Sister of Catherine Thorley, the governess at Down House between 1850 and 1856. Acted as governess to the Darwin children at Ilkley in October 1859.

Tollet, Georgina (1808–72). Daughter of George and Frances Tollet. A close friend of the Wedgwoods and Darwins. Edited the manuscript of *Origin*.

Treat, Mary (1830–1923). American botanist and entomologist. Née Davis; married Joseph Treat in 1863. Moved to Vineland, New Jersey, in 1868 to join the intellectual and agricultural community established by Charles Landis. Wrote many scientific and popular works on plants and insects from 1869. Separated from her husband in 1874 and supported herself by her writing and by collecting plant and insect specimens. Corresponded with Darwin, Asa Gray, C. V. Riley, August Forel, and Gustav Mayr. Advocate of the theory of natural selection. Her most notable research was on the anatomy and behaviour of harvesting ants, and on carnivorous plants.

Tyndall, John (1820–93). Irish physicist, lecturer, and populariser of science. Professor of natural philosophy, Royal Institution of Great Britain, 1853–87; professor of natural philosophy, Royal School of Mines, 1859–68; superintendent of the Royal Institution, 1867–87. An exponent of scientific naturalism; made controversial critiques of contemporary views on miracles and prayer. FRS 1852.

Vair, James (b. *c.* 1825 d. 1887). Scottish gardener. Sir Walter Scott's gardener at Abbotsford, Melrose, Scotland, and Lady Dorothy Fanny Nevill's head gardener at Dangstein, west Sussex, and Stillyans, Heathfield, east Sussex. A specialist in orchid growing.

Vaughan Williams, Margaret Susan (1843–1937). Daughter of Caroline Wedgwood and Josiah Wedgwood III. Darwin's niece. Married Arthur Charles Vaughan Williams in 1869. Mother of Ralph Vaughan Williams.

Wallace, Alfred Russel (1823–1913). Naturalist. Collector in the Amazon, 1848–52; in the Malay Archipelago, 1854–62. Independently formulated a theory of evolution by natural selection in 1858. Lecturer and author of works on protective coloration, mimicry, and zoogeography. President of the Land Nationalisation Society, 1881–1913. Wrote on socialism, spiritualism, and vaccination. FRS 1893.

Wedgwood, Caroline Sarah (1800–88). Darwin's sister. Married Josiah Wedgwood III, her cousin, in 1837.

Wedgwood, Charlotte. *See* Langton, Charlotte.

Wedgwood, Emma. *See* Darwin, Emma.

Wedgwood, Frances (Fanny) (1806–32). Daughter of Bessy and Josiah Wedgwood II; sister of Emma Darwin.

Wedgwood, Frances (Fanny) (1807–74). Daughter of the Rev. John Peploe Mosley of Rolleston, Staffordshire. Married Francis Wedgwood (Emma Darwin's brother) in 1832.

Wedgwood, Frances Emma Elizabeth (Fanny) (1800–89). Second child of James Mackintosh and Catherine Allen. Married Hensleigh Wedgwood in 1832.

Wedgwood, Frances Julia (Snow) (1833–1913). Novelist, biographer, historian, and literary critic. Daughter of Hensleigh and Frances Emma Elizabeth Wedgwood. Published two novels in her mid-twenties, one under the pseudonym Florence Dawson. Wrote book reviews and an article on the theological significance of *Origin*. Conducted an intense friendship with Robert Browning between 1863 and 1870. Published a study of John Wesley (1870), and helped Darwin with translations of Linnaeus in the 1870s. Published *The moral ideal: a historical study* (1888). Active in the anti-vivisection movement.

Wedgwood, Francis (Frank) (1800–88). Master-potter. Partner in the Wedgwood pottery works at Etruria, Staffordshire, until 1876. Emma Darwin's brother. Married Frances Mosley in 1832.

Wedgwood, Hensleigh (1803–91). Philologist. Emma Darwin's brother. Qualified as a barrister in 1828, but never practised. Fellow, Christ's College, Cambridge, 1829–30. Police magistrate at Lambeth, 1831–7; registrar of metropolitan carriages, 1838–49. An original member of the Philological Society, 1842. Published *A dictionary of English etymology* (1859–65). Married Frances Emma Elizabeth Mackintosh in 1832.

Wedgwood, Hope (1844–1935). Daughter of Hensleigh and Frances Emma Elizabeth Wedgwood. Second wife of Godfrey Wedgwood.

Wedgwood, Josiah I (1730–95). Master-potter. Founded the Wedgwood pottery works at Etruria, Staffordshire. Grandfather of Charles and Emma Darwin. Interested in experimental chemistry. Contributed several papers on the measurement of high temperatures to the Royal Society of London's *Philosophical Transactions*. Associated with scientists and scientific societies. FRS 1783.

Wedgwood, Josiah III (1795–1880). Master-potter. Partner in the Wedgwood pottery works at Etruria, Staffordshire, 1841–4; moved to Leith Hill Place, Surrey, in 1844. Emma Darwin's brother. Married Charles

Darwin's sister Caroline, his cousin, in 1837.

Wedgwood, Katherine Elizabeth Sophy (Sophy) (1842–1911). Daughter of Caroline Sarah Wedgwood and Josiah Wedgwood III. Darwin's niece.

Wedgwood, Katherine Euphemia (Effie). *See* Farrer, Katherine Euphemia (Effie).

Wedgwood, Lucy Caroline (1846–1919). Daughter of Caroline Sarah Wedgwood and Josiah Wedgwood III. Darwin's niece. Married Matthew James Harrison in 1874. Assisted Darwin in his work on botany, animal emotional expression, and worms.

Wedgwood, Sarah Elizabeth (Elizabeth) (1793–1880). Emma Darwin's sister. Resided at Maer Hall, Staffordshire, until 1847, then at The Ridge, Hartfield, Sussex, until 1862. Moved to London before settling in Down in 1868.

Wedgwood, Thomas (1771–1805). Son of Josiah Wedgwood I. Published researches on heat and light, 1791–2.

Whitby, Mary Anne Theresa (1784–1850). Landowner, antiquary, and artist. Of Newlands, Hampshire. Silk producer and author of *A manual for rearing silkworms in England* (1848).

White, James (1809/10–83). Merchant and politician. Married Mary Lind in 1833. Liberal MP for Plymouth, 1857–9; Brighton, 1860–74.

White, Mary (1815/6–83). Née Lind. Botanical artist. Married James White in 1833. Produced botanical watercolours on her trip through the United States to China in 1876.

Wilberforce, Samuel (1805–73). Clergyman. Bishop of Oxford, 1845–69. Famously spoke against *Origin of species* at the British Association meeting in 1860. FRS 1845.

Williams, Edward Hosier (d. 1844). Of Eaton Mascott, near Shrewsbury. Attorney in partnership with J. A. Powell and others at Lincoln's Inn. Marred Sarah Harriet Mostyn Owen in 1831.

Wolfe, Gould Anne (1824/5–85). Daughter of Henry Upton Ruxton RN of Ardee House, co. Louth, and Isabella, daughter of James Carlyle of Craddockstown, co. Kildare. Married Charles Wolfe, clergyman, in 1849. Lived in Dublin from 1867.

Wood, Thomas Fanning (1841–92). American medical editor and botanist. Resided in Wilmington, North Carolina; organised the state board of health. Founding editor, *North Carolina Medical Journal*, 1878. Wrote on the plants of North Carolina.

Wright, Chauncey (1830–75). American mathematician and philosopher. Calculator for the newly established *American ephemeris and nautical almanac*, for which he devised new methods of calculation, 1852–72. Recording secretary, American Academy of Arts and Sciences, Boston, 1863–70. Published the first of a series of philosophical essays in the *North American Review* in 1864.

Bibliography and further reading

Further reading

Abir-Am, Pnina G. and Outram, Dorinda, eds. 1987. *Intimate lives: women in science, 1789–1979*. New Brunswick: Rutgers University Press.
Biographical dictionary of women in science: pioneering lives from ancient times to the mid-20th century. Edited by Marilyn Ogilvie and Joy Harvey. 2 vols. New York and London: Routledge. 2000.
Browne, Janet. 1995. *Charles Darwin. Voyaging. Volume I of a biography*. New York: Alfred A. Knopf.
Browne, Janet. 2002. *Charles Darwin. The power of place. Volume II of a biography*. London: Pimlico.
Brück, Mary. 2009. *Women in early British and Irish astronomy*. Dordrecht: Springer. (Includes Thereza Dillwyn Llewelyn.)
Darwin, Bernard. 1955. *The world that Fred made: an autobiography*. London: Chatto & Windus.
Gianquitto, Tina. 2007. *'Good observers of nature'. American women and the scientific study of the natural world, 1820–1885*. Athens, Ga., and London: University of Georgia Press. (Includes Mary Treat.)
Hamlin, Kimberly A. 2015. *From Eve to evolution: Darwin, science, and women's rights in Gilded Age America*. Chicago: The University of Chicago Press.
Keynes, Randal. 2001. *Annie's box. Charles Darwin, his daughter and human evolution*. London: Fourth Estate.
Raverat, Gwen. 1952. *Period piece: a Victorian childhood*. London: Faber & Faber.
Richards, Evelleen. 1983. Darwin and the descent of woman. In *The wider domain of evolutionary thought*, edited by David Oldroyd and Ian Langham. Dordrecht and Boston, Mass.: D. Reidel Publishing Company.
———. 1997. Redrawing the boundaries: Darwinian science and Victorian women intellectuals. In *Victorian science in context*, edited by Bernard Lightman. Chicago: The University of Chicago Press.
Russett, Cynthia Eagle. 1989. *Sexual science: the Victorian construction of womanhood*. Cambridge, Mass., and London: Harvard University Press.
Sutherland, Gillian. 2006. *Faith, duty and the power of mind: the Cloughs and their circle, 1820–1960*. Cambridge: Cambridge University Press.
Walkowitz, Judith R. 1980. *Prostitution and Victorian society: women, class, and the state*. Cambridge: Cambridge University Press.
Wedgwood, Barbara and Wedgwood, Hensleigh. 1980. *The Wedgwood circle,*

1730–1897: four generations of a family and their friends. London: Studio Vista.

White, Paul. 2016. Darwin's home of science and the nature of domesticity. In *Domesticity and the making of modern science*, edited by Donald L. Opitz, Staffan Bergwik, and Brigitte Van Tiggelen. Basingstoke and New York: Palgrave Macmillan.

Bibliography

Anderson, Michael. 1990. The social implications of demographic change. In *The Cambridge social history of Britain, 1750–1950*, edited by F. M. L. Thompson, vol. 2. Cambridge: Cambridge University Press.

Barber, Mary Elizabeth. 1869. On the fertilization and dissemination of *Duvernoia adhatodoides*. [Read 15 April 1869.] *Journal of the Linnean Society (Botany)* 11 (1871): 469–72.

———. 1874. Notes on the peculiar habits and changes which take place in the larva and pupa of *Papilio nireus*. [Read 2 November 1874.] *Transactions of the Entomological Society of London* 22: 519–21.

Barlow, Nora, ed. 1958. *The autobiography of Charles Darwin 1809–1882. With original omissions restored.* Edited with appendix and notes by Nora Barlow. London: Collins.

Becker, Lydia Ernestine. 1864. *Botany for novices: a short outline of the natural system of classification of plants. By L.E.B.* London: Whittaker & Co.

———. 1869a. On the study of science by women. *Contemporary Review* 10 (1869): 386–404.

———. 1869b. On an alteration in the structure of *Lychnis dioica* observed in connection with the development of a parasitic fungus. *Journal of Botany* 7: 291–2.

Blackburn, Helen. 1902. *Women's suffrage: a record of the women's suffrage movement in the British Isles with biographical sketches of Miss Becker.* London: Williams & Norgate.

Boardman, Kay. 2004. Struggling for fame: Eliza Meteyard's principled career. In *Popular Victorian women writers*, edited by Kay Boardman and Shirley Jones. Manchester and New York: Manchester University Press.

Boole, Mary. 1883. *The message of psychic science to mothers and nurses.* London: Trubner & Co.

———. 1931. *Collected works.* With a preface by Ethel S. Dummer. 4 vols. London: C. W. Daniel.

Browne, Janet. 1995. *Charles Darwin. Voyaging. Volume I of a biography.* New York: Alfred A. Knopf.

Browne, Janet. 2002. *Charles Darwin. The power of place. Volume II of a biography.* London: Pimlico.

Buckley, Arabella Burton. 1876. *A short history of natural science and of the progress of discovery from the Greeks to the present day.* London: John Murray.

———. 1880. *Life and her children: glimpses of animal life from the amoeba to the insects.* London: Edward Stanford.

Buckley, Arabella Burton. 1883. *Winners in life's race, or the great backboned family.* London: Edward Stanford.

———. 1891. *Moral teachings of science.* London: Edward Stanford.

Burstyn, Joan N., ed. 1990. *Past and promise: lives of New Jersey women.* Metuchen, NJ, and London: The Scarecrow Press.

'Climbing plants': On the movements and habits of climbing plants. By Charles Darwin. [Read 2 February 1865.] *Journal of the Linnean Society (Botany)* 9 (1867): 1–118.

Cobbe, Frances Power. 1870. Hereditary piety. [Review of Francis Galton, *Hereditary genius*, 1869, and Prosper Despine, *Psychologie naturelle*, 1868.] *Theological Review* 7: 211–34.

———. 1871. Darwinism in morals. [Review of Charles Darwin's *Descent of man.*] *Theological Review* 8: 167–92.

[———.] 1872a. The consciousness of dogs. *Quarterly Review* 133: 419–51.

[———.] 1872b. Dogs whom I have met. *Cornhill Magazine* 26: 662–78.

———. 1904. *Life of Frances Power Cobbe as told by herself.* Posthumous edition. London: Swan Sonnenschein.

Colp, Ralph, Jr. 1972. Charles Darwin and Mrs. Whitby. *Bulletin of the New York Academy of Medicine* 2d ser. 48: 870–6.

Coral reefs 2d ed.: *The structure and distribution of coral reefs.* By Charles Darwin. Revised edition. London: Smith, Elder & Co. 1874.

Correspondence: *The correspondence of Charles Darwin.* Edited by Frederick Burkhardt *et al.* 23 vols to date. Cambridge: Cambridge University Press. 1985–.

Darwin, Francis. 1920. *Springtime and other essays.* London: John Murray.

Descent: *The descent of man, and selection in relation to sex.* By Charles Darwin. 2d edition. 2 vols. London: John Murray. 1871.

Descent 2d ed.: *The descent of man, and selection in relation to sex.* By Charles Darwin. London: John Murray. 1874.

Dixie, Florence. [1877.] *Abel avenged: a dramatic tragedy.* London: E. Moxon.

———. 1880. *Across Patagonia.* London: R. Bentley and son.

Earthworms: *The formation of vegetable mould, through the action of worms, with observations on their habits.* By Charles Darwin. London: John Murray. 1881.

Emma Darwin (1904): *Emma Darwin, wife of Charles Darwin. A century of family letters.* Edited by Henrietta Litchfield. 2 vols. Cambridge: privately printed by Cambridge University Press. 1904.

Expression: *The expression of the emotions in man and animals.* By Charles Darwin. London: John Murray. 1872.

Expression 2d ed.: *The expression of the emotions in man and animals.* By Charles Darwin. 2d edition. Edited by Francis Darwin. London: John Murray. 1890.

Herrick, Sophie Bledsoe. 1883. *The wonders of plant life under the microscope.* New York: G. P. Putnam's Sons.

———. 1885. *Chapters on plant life.* New York: Harper & Brothers.

———. 1887. Introductory essay on the genius of George Eliot. In *Essays and reviews of George Eliot not hitherto reprinted.* Boston: Aldine Book Publishing Company.

Herrick, Sophie Bledsoe. 1888. *The earth in past ages*. New York: Harper & Brothers.

Hirsch, Pam. 1998. *Barbara Leigh Smith Bodichon: feminist, artist, and rebel*. London: Chatto & Windus.

Hutchinson, Horace Gordon. 1914. *Life of Sir John Lubbock, Lord Avebury*. 2 vols. London: Macmillan.

Huxley, Thomas Henry. 1863. *On our knowledge of the causes of the phenomena of organic nature. Being six lectures to working men, delivered at the Museum of Practical Geology*. London: Robert Hardwicke.

———. 1869. On the physical basis of life. *Fortnightly Review* n.s. 5: 129–45.

Insectivorous plants. By Charles Darwin. London: John Murray. 1875.

Lewis, Alison M. 2000. Caroline Emelia Stephen (1834–1909) and Virginia Woolf (1882–1941): a Quaker influence on modern English literature, quakertheology.org/issue3-3.html (accessed 2 October 2015). (*Quaker Theology* 2: no. 2.)

Lightman, Bernard, ed. 2004. *Dictionary of nineteenth-century British scientists*. 4 vols. Bristol: Thoemmes Continuum.

Lubbock, Ellen Frances. 1864. The ancient shell-mounds of Denmark. In *Vacation tourists and notes of travel in 1862–3*, edited by Francis Galton. London and Cambridge: Macmillan and Co.

———. 1875. [Review of Charles Darwin's *Insectivorous plants*.] *Academy*, 24 July 1875, pp. 93–4.

Lyell, Katharine Murray. 1870. *A geographical handbook of all the known ferns with tables to show their distribution*. London: John Murray.

———, ed. 1881. *Life, letters and journals of Sir Charles Lyell, Bart*. 2 vols. London: John Murray.

———, ed. 1890. *Memoir of Leonard Horner … consisting of letters to his family and from some of his friends*. 2 vols. London: privately printed.

Martineau, Harriet. 1864. Middle-class education in England. Girls. *Cornhill Magazine* 10: 549–68.

Meteyard, Eliza. 1845. *Struggles for fame*. London.

———. 1865–6. *The life of Josiah Wedgwood from his private correspondence and family papers … with an introductory sketch of the art of pottery in England*. 2 vols. London: Hurst & Blackett.

Michelet, Athénaïs. 1904. *Les chats*. Paris: Ernest Flammarion.

Miller, Hugh. 1849. *Footprints of the Creator: or, the Asterolepis of Stromness*. London: Johnstone and Hunter.

Morris, Margaretta Hare. 1840. On the *Cecidomyia destructor*, or Hessian fly. [Read 2 October 1840.] *Transactions of the American Philosophical Society* n.s. 8 (1843): 49–52.

———. 1841. Observations on the development of the Hessian fly. [Read 10 August 1841.] *Proceedings of the Academy of Natural Sciences of Philadelphia* 1 (1841–3): 66–8.

———. 1846. [On the larvae of *Cicada septemdecim* preying on the roots of fruit trees.] [Read 15 December 1846.] *Proceedings of the Academy of Natural Sciences of Philadelphia* 3 (1846–7): 132–4.

———. 1849. [On *Cecidomyia culmicola*.] [Read 21 August 1849.] *Proceedings of the Academy of Natural Sciences of Philadelphia* 4 (1848–9): 194.

Morris, Margaretta Hare. 1851. [On the seventeen years' locust.] [Read 1 October 1851.] *Proceedings of the Boston Society of Natural History* 4 (1851–4): 110–11.

Nash, Louisa A'hmuty. 1890. Some memories of Charles Darwin. *Overland Monthly* 16: 404–8.

Nevill, Ralph. 1919. *The life & letters of Lady Dorothy Nevill.* London: Methuen & Co.

North, Marianne. 1893. *Recollections of a happy life, being the autobiography of Marianne North.* 2d edition. 2 vols. Edited by Janet Catherine Symonds. London and New York: Macmillan and Co.

ODNB: Oxford dictionary of national biography: from the earliest times to the year 2000. (Revised edition.) Edited by H. C. G. Matthew and Brian Harrison. 60 vols. and index. Oxford: Oxford University Press. 2004.

Orchids: On the various contrivances by which British and foreign orchids are fertilised by insects, and on the good effects of intercrossing. By Charles Darwin. London: John Murray. 1862.

Origin: On the origin of species by means of natural selection, or the preservation of favoured races in the struggle for life. By Charles Darwin. London: John Murray. 1859.

Pauly, Daniel. 2004. *Darwin's fishes. An encyclopedia of ichthyology, ecology, and evolution.* Cambridge: Cambridge University Press.

Pfeiffer, Emily. 1873. *Gerard's monument: and other poems.* London: Trubner.

'Recollections': Recollections of the development of my mind and character. By Charles Darwin. In *Evolutionary writings,* edited by James A. Secord. Oxford: Oxford University Press. 2008.

Richardson, Benjamin Ward. 1876. Abstract report to 'Nature' on experimentation on animals for the advance of practical medicine. *Nature,* 15 June 1876, pp. 149–52; 22 June 1876, pp. 170–2; 29 June 1876, pp. 197–9; 20 July 1876, pp. 250–2; 3 August 1876, pp. 289–91; 17 August 1876, pp. 339–41; 31 August 1876, pp. 369–72.

Shepherd, John A. 1980. *Lawson Tait: the rebellious surgeon (1845–1899).* Lawrence, Kansas: Coronado Press.

Smith, Mavis E., ed. 2008. *Ellen Tollet of Betley Hall. Journals and letters from 1835.* Crewe: the author.

Somerville, Mary. 1869. *On molecular and microscopic science.* 2 vols. London: John Murray.

——. 1877. *The connexion of the physical sciences.* 10th edition, revised and corrected by Arabella Burton Buckley. London: John Murray.

'Three forms of *Lythrum salicaria*': On the sexual relations of the three forms of *Lythrum salicaria.* By Charles Darwin. [Read 16 June 1864.] *Journal of the Linnean Society (Botany)* 8 (1865): 169–96.

[Tollet, Georgina.] 1886. *Country conversations.* London: Whiting & Co.

Treat, Mary. 1871. *Drosera* (sundew) as a fly-catcher. *American Journal of Science and Arts* 3d ser. 2: 463.

——. 1873a. Controlling sex in butterflies. *American Naturalist* 7: 129–32.

——. 1873b. Observations on the sundew. *American Naturalist* 7: 705–8.

Treat, Mary. 1875. Plants that eat animals. *American Naturalist* 9: 658–62.

———. 1876a. Is the valve of Utricularia sensitive? *Harper's New Monthly Magazine* 52: 382–7.

———. 1876b. Carnivorous plants of Florida. *Harper's New Monthly Magazine* 53: 546–8, 710–14.

Variation: The variation of animals and plants under domestication. By Charles Darwin. 2 vols. London: John Murray. 1868.

Wedgwood, Barbara and Wedgwood, Hensleigh. 1980. *The Wedgwood circle, 1730–1897: four generations of a family and their friends.* London: Studio Vista.

[Wedgwood, Frances Julia.] 1860–1. The boundaries of science, a dialogue. *Macmillan's Magazine* 2 (1860): 134–8; 4 (1861): 237–47.

[———.] 1873. [Review of *The fair haven*, by Samuel Butler.] *Spectator*, 15 November 1873, pp. 1440–2.

[———.] 1881. The moral influence of George Eliot. By one who knew her. *Contemporary Review* 39: 173–85.

Whitby, Mary Anne Theresa. 1848. *A manual for rearing silkworms in England: with a brief notice on the cultivation of the mulberry tree.* London.

Index

Page numbers in bold type refer to letters to or from, italic numerals refer to illustrations.